THE COUNTRYSIDE IDEAL

One of the contradictions of modern urban civilisation is the persistence of a nostalgia for rural life and landscape which has raised the countryside to an idealised status. So far the discussion of this phenomenon has been restricted to the relatively narrow perspectives of literary and intellectual history. This book broadens the analysis of the countryside ideal by exploring the relationships between its cultural origins and its manifestation in contemporary landscapes.

The Countryside Ideal examines the main historical processes and ideas underlying the continuing attachment to the countryside, and how these have influenced popular values and lifestyles, defined attitudes to nature, country life and landscape, and affected the development of rural and urban landscapes. The cultural geographical framework recognises the particular strength of the countryside ideal in Anglo-American culture, and explores the similarities and differences in its British and North American expression.

This book draws together diverse images of landscape to explore this preoccupation with place, culture and representation in Anglo-American society.

Michael Bunce is Associate Professor of Geography at Scarborough College, University of Toronto, Canada.

THE COUNTRYSIDE IDEAL

Anglo-American Images of Landscape

Michael Bunce

London and New York

First published 1994
by Routledge
11 New Fetter Lane London EC4P 4EE

Simultaneously published in the USA and Canada
by Routledge
29 West 35th Street, New York NY 10001

© Michael Bunce

Phototypeset in Garamond by Intype, London

Printed and bound in Great Britain by
Biddles Ltd, Guildford and King's Lynn

British Library Cataloguing in Publication Data
A catalogue record for this book is available from the British Library.

Library of Congress Cataloging in Publication Data
Bunce, M. F.
The countryside ideal : Anglo-American images of landscape / Michael Bunce
p. cm.
Includes bibliographical references (p.) and index.
1. Great Britain–Historical geography. 2. United States–Historical geography. 3.
Landscape–Great Britain–Historiography. 4. Landscape–United
States–Historiography. 5. Ideals (Aesthetics)
I. Title
DA600.B79 1994
911'.41–dc20
93–33662
CIP

ISBN 0–415–10434–3
0–415–10435–1 (pbk)

To my parents

CONTENTS

PLATES

PREFACE

For almost twenty years now I have been an observer of the changing rural scene. Like most of my colleagues in this field, my research has been preoccupied by the impact of urban-based change on rural areas. What has increasingly struck me about this perspective is its inherently protectionist view of the countryside. Rural change is invariably seen as an invasive process which disrupts traditional rural communities and degrades rural landscapes. But, for academics and especially social scientists who so frequently claim the high ground of objectivity, why should rural change – or as current jargon now calls it 'restructuring' – be regarded in such a negative light? Perhaps it could be explained by the dominance of the problem paradigm in the social sciences; all processes create problems and it is the job of the social scientist and social-scientist-turned-planner to identify and solve them. Perhaps it is the natural bias of those who label themselves experts in *rural* studies which have led them to view the subject from only a rural perspective. But perhaps it is something more general than this, something which lies outside the frameworks of academic inquiry. The more I have considered this question over the years the more I have sensed the influence of a broader valuation of the countryside in urban society in general. As largely urban-based analysts of rural change we are incapable of separating ourselves from this. Moreover, because this sentiment towards the countryside has had a great effect on the nature of rural change itself, I have come also to recognise the need to incorporate it more thoroughly into our analyses.

Some years ago, therefore, I began to investigate what the countryside means to modern urban society and how the values that flow from this have influenced rural life and landscape. An initial survey of the literature which might have helped me in my inquiry was frustrating, for it was restricted to a few passing observations by those who were writing about rural change, to the literary analysis of the culture of pastoral sentiment by English scholars, or to examinations of the particular ideologies which have influenced rural sentiment. A broader and more integrated perspective was clearly needed. As a geographer I was interested in exploring the

xi

relationships between values and landscape change, between, that is, how we perceive and how we treat the countryside. As someone who has lived and worked in both Britain and North America, and who has become aware of the linkages of Anglo-American culture I was interested also in examining the topic from a transatlantic perspective.

And so the idea of this book was born way back in the autumn of 1983 in a conversation with Roger Jones, then editor with George Allen and Unwin. This led to an outline and then a contract, which, in the true tradition of academic writing, stretched past many deadlines. Ten years of peripatetic research and writing, and a couple of mergers of publishing companies later, what I hope is a unique and useful book has finally been published. Understandably, over such a long period of gestation many of my original, and, in retrospect, often naive ideas have changed. The result is a study which, hopefully, reveals the complexities of the subject which I discovered in its writing. I have aimed the book at an academic, professional and general readership; at those who study, plan and appreciate the countryside in its broadest sense. To satisfy such a wide constituency is a tall order, but hopefully there is something in here to at least whet the appetite of every reader for this fascinating topic. The countryside ideal is very much alive as we approach the end of the millennium. The better our understanding of its origins, its development and its consequences, the better placed we shall be to understand and manage our landscapes.

ACKNOWLEDGEMENTS

Of the various family, friends and colleagues who have assisted me directly and indirectly with this project, I wish to thank first and foremost my wife, Barbara, for her support. Even though she has had her own demanding career to contend with, she has always been there to help with advice, editing and, above all, constant encouragement. Thanks are due to my son, Jonathan, for proof-reading the final manuscript, to my daughter Susannah, for our interesting historical conversations, and to my father, George Bunce, and my brother, Peter Bunce, for taking the photographs of Wilton and Southampton Common at short notice. Beyond this family circle, I would like to thank my colleague Ted Relph for our discussions about landscapes and life in general, but especially for his advice on Chapter 5. Chris Bryant, Kim England, Jim Lemon, Jeanne Maurer, Mike Troughton and Jerry Walker have also contributed ideas and sources. I owe special debts of gratitude to David Harford of the Scarborough Graphics and Photography Department and to Audrey Glasbergen for her assistance with copyright correspondence. Finally, I would like to acknowledge Roger Jones for his original faith in this project, and Tristan Palmer for his excellent editorial work since Routledge acquired the rights.

Acknowledgements are due for permission to reproduce the following material:
The Biltmore Company for Plate 3.2 (The Biltmore Company has granted Michael Bunce permission to use its trademark in connection with *The Countryside Ideal*; The British Tourist Authority for Plate 4.5; British Waterways for Plate 5.10; Brunner Mond and Co. for Plate 2.7; Faber and Faber Ltd for Plate 3.4; David Harford for Plate 3.3; The Ford Motor Company for Plate 4.3; The Hamilton Group for Plate 2.5; The Imperial War Museum for Plate 1.2; The London Transport Museum for Plate 5.8; The Museum of London for Plate 5.1; The National Parks Service for Plate 6.3; The National Trust for Historic Preservation and Maguire/Reeder for Plate 6.4; Old Sturbridge Village and Robert S. Arnold for Plate 4.9; Priory Gallery for Plate 2.3; Routledge for Plate 6.1; Trustees of the British

Museum for Plate 1.1; Frederick Warne and Co. for Plate 2.4. Plate 2.1 is reproduced by permission of the Metropolitan Museum of Art, Gift in memory of Jonathan Sturges by his children, 1895 (95.13.3). Plate 2.2 is reproduced by courtesy of the Trustees, The National Gallery, London. Plate 2.6: copyright 1901, Meredith Corporation. All Rights Reserved. Reprinted from the *Ladies Home Journal* magazine with permission of SmithKline Beecham. Plate 3.6 is reprinted by permission from the April 1987 issue of *Country Living*, © 1987 by the Hearst Corporation. Photography by Elyse Lewin. Plate 4.4 is reprinted with permission of the American Automobile Manufacturer's Association. Plate 5.3 is reproduced courtesy of the Frances Loeb Library, Graduate School of Design, Harvard University. Plate 5.6 is reproduced with the kind permission of the Town and Country Planning Association.

Every effort has been made to obtain permission to reproduce copyright material. If any proper acknowledgement has not been made the publishers invite the copyright holder to inform them of the oversight.

INTRODUCTION

'God made the country, and man made the town.' Although they were written over two hundred years ago, these much-quoted words of William Cowper still seem to reflect popular sentiment. As we approach the end of the twentieth century, one of the many contradictions of what is fashionably described as post-modern society, is the continued, even growing nostalgia for the countryside and an abiding ambivalence towards the city. In a great variety of forms – in popular literature and the electronic media, in country homes and weekend cottages, in hobby-farms and back-to-the-land communities, in nature trails and wilderness trips, in heritage villages and country parks, in suburban design and commercial imagery, in conservation movements and planning policy – the countryside indeed appears to retain the reverential associations bestowed upon it by Cowper.

Although it is one of the enduring paradoxes of modern western civilisation, it is also a phenomenon which reaches back as far as classical times, when the emergence of city life produced the first sense of the distinctiveness of rural and urban worlds. Since then the country and the city have acquired a range of contrasting associations. These have often cast the countryside in a negative light, comparing it unfavourably to the sophistication and power of the city. Yet nostalgia for pastoral golden ages, as Raymond Williams's seminal analysis of literary responses to the country and the city so eloquently argued, is as old as civilisation itself (Williams 1973). And whenever urban civilisations have reached their zenith the pendulum, as Tuan (1974) has observed, appears to have swung especially strongly in favour of the country. No urban epoch has exhibited this more than our own. As the modern western urban–industrial system has tightened its grip on life and landscape, sentiment towards the countryside seems to have reached idealistic proportions, acquiring almost mythological status in our mental view of the world and at the same time becoming increasingly valued as a tangible alternative to urban life.

Many writers have dismissed this as simplistic urban sentimentalism and escapism. Yet there is considerably more to the idealisation of the countryside than this. In the first place it is deeply entrenched in our value system

1

– ideologically, psychologically and culturally. It contributes significantly to the determination of our relationships with nature and society. And it is also a powerful influence on the way we view and treat our cultural landscapes, both urban and rural.

At its most profound level, the affection for the countryside may reflect fundamental human values and psychological needs which can be traced to a basic human desire for harmony with land and nature, for a sense of community and place and for simplicity of lifestyle. With the rise of urban–industrial society these needs have been magnified and projected on to a countryside redefined as the symbolic antithesis of the city; a place for reconnecting to natural processes and ancestral roots. Yet the idealisation of the countryside is intricately bound up in the development of modern urban civilisation. It is a product of three centuries or so of changes which have accentuated the cleavage between country and city through the transformation of their respective landscapes and the redefinition of human relationships with land, nature and community. With this has come a level of affluence, mobility and education from which has developed a broadening of lifestyle choices and an increasing influence over the quality of the human environment. The attraction of the countryside, therefore, cannot be explained solely in terms of elemental human needs. It is also a cultural construct and a social ideal, forged by the historical processes of a metropolitan-dominated society. From these have emerged the mix of ideology and values, myth and stereotype, image and perception, as well as lived experience which has sustained the ideal.

Much has been written about the idealisation of the countryside. However, this has been dominated by the relatively narrow and specialised perspectives of literary and intellectual history. Our attitudes to the countryside therefore tend to be understood largely in terms of their literary and artistic expression and in the articulation of ideas about the country and the city by great thinkers. This, of course, has yielded a rich literature, which, as I demonstrate in the early chapters of this book, is central to the discussion of the countryside ideal. Yet it is an ideal which has grown beyond its cultural and philosophical origins into the realms of popular and tangible expression in the actual landscapes and living spaces of modern society.

My aim in this book, therefore, is to broaden the analysis of the countryside ideal by exploring the relationships between its cultural origins and its manifestation in contemporary landscapes. The approach is thus fairly, if not squarely, located within the general realm of cultural geography, in which the historical changes in society are seen as symbolised in everyday landscapes. The countryside ideal is therefore interpreted in terms of the tripartite relationship between the historical forces through which it has been formed, the symbolic meanings which it has given to particular landscapes and the reinforcement of these meanings in the formation of

2

cultural landscapes themselves (for a full discussion of this approach see Cosgrove 1984; Meinig 1979; and Relph 1981). What are the main historical processes and ideas which lie behind the continuing attachment to the countryside? How have these been nurtured within Anglo-American culture? How have they influenced popular values and lifestyles? How have they defined attitudes to nature, rural life and landscape? How have they affected the development of rural and urban landscapes? What are the consequences for society and environment? These are the central questions of the chapters that follow.

This is therefore both an historical and a geographical analysis. Its historical component recognises that the countryside ideal has emerged with the evolution of industrial society, in the development of the myths and values that have come to be associated with the countryside as well as with the evolution of its actual use and treatment. It recognises, in other words, that this is not just a recent phenomenon but that it is deeply entrenched in our culture and our landscape history. The geographical emphasis of the book recognises the profound affect that the rise of the countryside ideal has had on natural and cultural landscapes, especially on the use and appearance of rural and wilderness areas but also on the landscapes of cities and suburbs.

Although I refer briefly to classical antecedents, the historical scope of the book takes us back to the early eighteenth century, covering the roughly three centuries of the rise of modern western metropolitan civilisation as we know it. The geographical context is that of Britain, primarily England, and North America, mainly the USA. In part this reflects my own cultural and academic experience and the need for a discussion which is manageable in a relatively short book. But it also reflects the strength and distinctiveness of what can be called, for want of a better term, the Anglo-American version of the countryside ideal. Countryside, as I have suggested, is a culturally constructed term. It is also very much an English term which reflects a peculiarly national obsession with the countryside as an aesthetic and a social ideal. Much of this cultural baggage has accompanied its transfer across the Atlantic, and so there are important links between English and North American attitudes. At the same time, there are significant differences, born largely of different environmental circumstances, settlement histories and social systems, which have made the American countryside ideal more diffuse. For largely geographical reasons, a specifically Canadian countryside ideal appears to be poorly developed. It has certainly received little academic attention. The North American references in this book therefore are mainly to the USA.

So far I have avoided defining countryside. It is, in fact, a nebulous term, which defies precise definition. That it first came into common usage to describe the gentrified rural landscapes of early eighteenth-century England reveals its thoroughly English origins. Indeed there would be little disagreement in Britain, and especially England, as to the meaning of 'countryside', referring as it does to the aesthetic and amenity qualities of a universally domesticated rural landscape, and especially to the landscape of agricultural enclosure. That this has become a nostalgic symbol of English national identity poses some difficulties in transferring the word 'countryside' to the North American context. Although, with the recent rise of public interest in the conservation of settled rural landscapes, it is beginning to creep into more general use, historically 'countryside' has not meant much to either Americans or Canadians. For reasons which I explore in Chapters 1 and 6, they have tended to downplay the value of agricultural landscapes, preferring instead to turn to more natural settings for aesthetic and amenity appreciation, and to historical artifacts for nostalgic satisfaction. In this book, then, 'countryside' is a liberally-used and intentionally vague term, at once cultural and geographical, which encompasses the whole range of environments from wilderness to market town that constitute the objects of the countryside ideal.

Clearly there is lot of ground to cover in exploring this topic. In setting out to discuss its historical and ideological origins and its manifestations in the treatment of landscape in a comparative Anglo-American survey, I recognise that I have taken on a huge topic. However, in the chapters that follow, by combining overview with specific example, I attempt to provide an impression of the dimensions and the impacts of an ideal which has become a remarkably persistent feature of modern society.

1
THE MAKING OF AN IDEAL

Like all ideals, the modern countryside ideal is a creation of the society within which it has developed. It has been fashioned from the combination of historical processes and cultural values of the past three centuries, and must therefore be understood in terms of the evolving experience of metropolitan life. This chapter explores how pro-countryside sentiment has emerged with the rise of urban-industrialism, firstly through the transformation of rural–urban relationships and the respective economies and landscapes of country and city; and secondly through the development of philosophical, aesthetic and social responses to the urbanisation process itself.

THE RISE OF URBANISM

New cities, new countrysides

Modern sentiment for the countryside is the latest version of an ancient theme. From the fragmentary writings which survive from the city-states of Mesopotamia and Egypt we can identify a nostalgic interest in the differences between agricultural and non-agricultural people and a sense of urban separation from the natural world (Sorokin *et. al.* 1965). In ancient China agriculture was regarded as a superior activity to the commerce of cities. Although, according to Tuan (1974), Chinese scholarship wavered for two millennia between the attractions of city and country, Chinese society was obsessed by a recurrent fear that it would lose sight of its agrarian foundations. In classical Greece the dominance of cities in the Alexandrian Age produced a strong reaction against urban sophistication and a nostalgia for agrarian rusticity. Sentiment for the countryside became a significant philosophical ideal as well as a popular literary device. Hesiod and Xenophon's regressive notions of an agrarian Golden Age and the idylls of Theocritus and the other rustic poets established the countryside as the metaphor for a pastoral other-world (Williams 1973). In Augustan

Rome, literary pastoralism was matched by a growing interest in the countryside as a place of relaxation and pleasure. The poetic reaction against the city of the leading figures of the Roman pastoral tradition, Virgil and Horace, was derived as much from their experience as gentlemen farmers as from their imagination. Their Arcadian vision must be viewed in the context of their ability to retreat to their country estates (White 1977). This was a reaction to the increasingly crowded decadence of Rome which, with the imaginative pastoralism of literature, established the first broad tradition of urban perception of the countryside as an amenity.

While classical civilisation's attitudes to the countryside reveal how educated society has responded over the longer span of history to the process of urbanisation, its links to the modern countryside ideal are tenuous. As we shall see in the next chapter, they exist primarily through neo-classical revivalism in art and literature rather than through historical continuity. The period which followed the demise of the Roman Empire marked a real break in the evolution of urban civilisation. Most of the European landscape remained thoroughly rural as well as being extensively unsettled. Nor did the emergence of the medieval city bring renewed sentiment in favour of the countryside. On the contrary, it was viewed by medieval urban society with a sense of urbane and scholarly detachment, which found expression in a general literary disdain for country life (Tuan 1974). The fact, too, that most of the population and much of the political and economic power of the times still resided in rural areas, and that towns were largely small and isolated places ensured that the notion of countryside as a pastoral contrast to urbanism could have little meaning.

As cities grew in size, power and frequency in the late Middle Ages, the distinction between rural and urban began to acquire renewed significance. The fifteenth century in Italy saw the rise of modern Europe's first large cities. Centres of learning, progress and wealth, Venice, Florence and Sienna were culturally aloof from, and, to a degree, economically independent of the surrounding countryside. In design, too, they were uncompromising in their artificiality and in the clarity of their boundaries. Renaissance scholars continued the medieval disinterest in rural life, but the Renaissance did see the revival of classical pastoralism. The differences between city and country began again to be the metaphorical context for poetic and artistic images of artificiality and nature, while affluent and educated urbanites found pleasure in the rural landscape and the acquisition of country houses (Newton 1971). By the sixteenth century, these values were beginning to appear in English society. Indeed, it is in Tudor and Elizabethan England, the period of the so-called English Renaissance, that the conditions for the subsequent development of the Anglo-American countryside ideal begin to appear. Although England was still very much a rural nation, it was a nation undergoing fundamental changes. It was an age, especially towards the end of the century, marked by a great upsurge of capital investment in industry, agriculture and, above all, land (Merrington 1976). Not only did this hasten the breakdown of the feudal-ism which had governed rural society and economy for several centuries, but it also marked the beginnings of a transformation in the character of and the values associated with the English countryside.

At the heart of this was a rapid increase in investment in land and property. It was a period of great opportunism in the land market and of upward mobility in society which placed increasing amounts of land in the hands of country squires and yeoman farmers (Butlin 1982). Through-out the seventeenth century the influence of the landed gentry steadily increased. It has been estimated that by 1640 the middle and lesser gentry held about half of the land in England and Wales and the larger counties supported hundreds of gentry families (Cooper 1978). As the century progressed, an active land market and a growing number of marriage settlements concentrated ownership into fewer and larger estates. By 1700 large landowners controlled between 70 and 75 per cent of the cultivated land in England and Wales, while a century later their power is estimated to have extended to as much as 85 per cent of this land (Butlin 1982).

This was the major catalyst for the rise of capitalist modes of agricultural production, for what emerged from this process was a classic landlord–capitalist/tenant–wage labour structure (Merrington 1976). The small free-holders, copyholders and cottagers which made up the independent peasan-try were gradually eased out by growing numbers of tenant farmers and a new class of rural proletariat. The transition from low to high rates of population growth at the end of the sixteenth century provided a stimulus to the steady growth of agrarian capitalism, which in turn became the basis for a transformation of agricultural production from largely subsistent to commercial objectives. New crops and new methods of livestock pro-duction foreshadowed the coming revolution in agricultural technology. Enclosure of the open field and the conversion of arable land to pasture were central issues in this process, for not only did this establish the conditions for further capitalisation of agriculture but it also directly threatened the security of the peasantry. Although displacement of the rural population by enclosure and conversion was not as widespread as observers of the time complained, it is clear that, from the 1580s onwards there was a sharp increase in the mobility of the rural population and a drift from agricultural to non-agricultural occupations (Chambers and Mingay 1966).

In the changes that occurred in seventeenth-century England, then, the essential pre-conditions for subsequent shifts in rural–urban relationships and hence in perceptions of the countryside were set. Central to this was a new social hierarchy of landlord, tenant farmer and labourer together with a new class of rural entrepreneurs to serve and exploit the agrarian economy. The controlling influence on the countryside, of course, was held by the landed gentry and the aristocracy. They occupied an influential position in both town and country, supporting their urban and industrial enterprises with the economic resources and political power of their country estates and in turn reinvesting the profits of commerce and industry in the countryside (Jones 1968). They also exercised enormous influence over rural society in their own counties and parishes. The country estate was a mark of prestige and a symbol of dynastic control. What was sought was that 'mixture of enterprise, interest, rural pleasure and social prestige which the successful cultivation of a country estate brings' (Humphreys 1964: 24).

As estates spread across whole counties, the seventeenth- and eighteenth-century countryside came to be viewed, by cultured society at least, increasingly through the filter of the social order and gentrified lifestyles which these estates sustained. It is in early eighteenth-century England, in fact, that the word 'countryside' first comes into common usage as a reference to the amenity value of the rural landscape. For the landed classes it was associated with the sporting pleasures of hunting, shooting and fishing, and with a life of genteel ease in carefully landscaped surroundings. Fashionable country seats became part of the regular itinerary of a growing fad for touring the countryside (Ousby 1990). One of the more notable travellers was Daniel Defoe who, in his *Tour through the Whole Island of Great Britain* published in the 1720s extolled the virtues of a landscape which 'shines with a lustre not to be described' (in Humphreys 1964). The association by the landed classes of the countryside with aesthetic and recreational pleasure was also reflected in the popularity of pastoral literature and landscape painting which flourished largely through the patronage of the country estates which served as their favoured settings. In the England of the mid-eighteenth century, in the predominantly provincial Georgian England built upon two centuries of gentrification, with its wealth firmly planted in rural land and market towns, its progress measured by the spirit of improvement in agriculture and scenery and its society gauged through the ordered structure of parish and county, the idealisation of the countryside came largely from those who directly enjoyed its benefits, rather than from any significantly urban-based nostalgia.

Beneath this apparently serene rural order, however, lurked the forces of fundamental changes in economy and society. By the 1750s Britain was a predominantly market economy. It was the world's leading trading nation, drawing its wealth from an expanding overseas market and its growing colonial possessions. This supported a flourishing domestic economy founded on commercial agriculture, small-scale industry and mining, all hooked into a network of market towns with London as the supreme centre of commerce. Although demographically still predominantly a rural nation, it was one in which virtually all vestiges of a peasant economy had disappeared. In its place had emerged not only capitalist systems of agricultural production, but also an increasingly diversified rural economy based upon the processing and trading of agricultural commodities as well as the local manufacturing of agricultural implements, household articles and other trappings of an increasingly affluent society. In this economy lay the pre-conditions for the industrial revolution: the creation of domestic capital, innovations in transportation necessitated by the growth in internal trade, the production of food as well as consumer and capital goods (such as textiles and building materials), and the establishment of a growing domestic market organised through a monetary system of exchange (Hobsbawm 1968). Above all it created out of a growing body of landless labourers the proletariat which was necessary for the establishment of industrial capitalism (Merrington 1976). For much of the eighteenth century this proletariat was sustained by the growth of the rural economy itself. Yet, with the general increase in population growth which came after 1750, the ability of this economy to absorb an expanding labouring population steadily diminished (Mathias 1969). The consequences were widespread rural poverty and distress towards the end of the century which, despite the settlement laws which were supposed to tie people to their home parishes for the purposes of poor relief, led to a significant increase in the mobility of the rural proletariat (Dunford and Perrins 1983).

Far from being imposed upon the countryside by the unseen and unwelcome hand of technology, then, the urban–industrial system emerged to a great extent out of the changing structure of the rural economy itself. Of course, the direct catalysts for the transformation of Britain from a rural and agricultural nation to a predominantly urban and industrial one were the technological innovations and capital investments which created the industrial system. The huge labour demands of factory production and mining readily absorbed the surplus rural population. By the middle of the nineteenth century, over 40 per cent of the working population was engaged in mining, manufacturing and construction and a further 14 per cent in trade and transportation. At the same time the proportion engaged in agriculture and related occupations dropped from an estimated 75 per cent of the population in 1750 (Mathias 1969), to just over 21 per cent in 1851 (Pollard and Crossley 1968).

By the middle of the nineteenth century Britain had become a thoroughly industrial nation, leading the world in manufacturing and trade. In the process it became the first predominantly urbanised nation. With the industrial revolution's dependence on large concentrations of labour and materials came phenomenal rates of increase in both the size and the number of large cities. Up to the mid-eighteenth century, only London and Edinburgh had more than 50,000 inhabitants (Hobsbawm 1968) and, with a population of over 600,000, only London approached the scale and the conditions of the modern city. By the beginning of the nineteenth century there were eight cities with a population of over 50,000. At mid-century this number had grown to 29, including nine of over 100,000. By century's end Britain had 30 cities with populations of more than 100,000 (Weber 1899). Urban rates of population growth exceeded 20 per cent in each decade between 1801 and 1851, and reached almost 30 per cent a decade between 1811 and 1831. What is particularly notable about British urbanisation, however, is how comprehensively the demographic emphasis shifted from rural to urban during the nineteenth century. As the urban share increased from a third of the national total at the beginning of the century to over three-quarters by the end, whole rural areas experienced absolute depopulation (Saville 1957).

Although the drift from the land continued until well into this century, the urbanisation and industrialisation of Britain was almost complete by 1900. Across the Atlantic, of course, this transformation began and was completed somewhat later. Yet the European settlement of North America emerged out of the very changes which have been described above. By the time of American independence commercial, if not industrial, cities were already the established model for economic growth. Jefferson could dream of an agrarian republic, but the reality was that America had to compete with an increasingly urban and industrial Europe. And so, although the continued spread of settlement westwards guaranteed high levels of rural and agricultural population growth throughout the nineteenth century, urbanisation took hold equally strongly in the east. Most of this growth did not begin until after 1830 (Hahn and Prude 1985). Over the next hundred years, however, both the USA and Canada became predominantly urban nations. In the USA, the number of cities of over 100,000 grew steadily from five in 1830 to almost 100 by the time rates of urbanisation began to level off in 1930, while the overall urban population grew from 15 per cent of the national total to over 60 per cent (Monkonnen 1988). In Canada, although clearly the number of large cities which could develop was far smaller, the general rates of urban growth closely mirrored those in the USA (Stone 1967). In both countries urbanisation was closely linked to national growth, and thus to an expanding agrarian economy. Indeed until after the Civil War, it was trade rather than manufacturing which supported the growth of larger cities (Glaab and Brown 1976). However, the rapid urban growth that occurred after 1860 resulted from a steady increase in the movement of people from Europe and from the North American countryside itself into an expanding industrial system.

Urban society and the countryside

Viewed in simple terms, the idealisation of the countryside was an inevitable consequence of the urban–industrial revolution. Certainly it involved rapid and profound changes in society, economy and landscape, a process which at its nineteenth-century height ensured virtually continuous turmoil and instability. What better conditions could there be for the development of nostalgia for the countryside? Yet the countryside ideal cannot be explained simply as a nostalgic reaction to urbanisation. Rather we must see it as an ideal which has emerged from the very nature of modern urbanism itself. Urbanisation established four basic conditions for the nurturing of the countryside ideal. It produced the social structures and experiences within which attitudes towards the country and the city could develop. It created a political economy which redefined rural–urban relationships. It sustained the intellectual and cultural climate in which ideas about the country and the city could flourish. And, finally, it forged the landscapes and living environments around which differential values have formed.

With the massive shifts of people from agricultural to industrial occupations and from rural to urban living came fundamental changes in the general structure of society. As we have seen, the traditional and largely static hierarchical relationships which are conventionally associated with pre-industrial society were already beginning to break down in the period leading up to the industrial revolution. And in colonial North America the very process of agricultural settlement was founded on a rejection of the old European rural order. Yet as industrial systems of production and employment became dominant and thus drew an ever larger proportion of the population into cities, the breakdown of the old social order was rapid and complete. With urban capitalism society was increasingly organised along horizontal, rather than hierarchical lines, within homogenised groupings which divided people into new and simplified class categories.

For the proletarian masses the shift from country to city, and especially from agricultural to industrial employment destroyed a whole way of life. This has been widely studied and discussed by historians (notably Hobsbawm 1968 and Thompson 1968). Briefly summarised, it involved the replacement of the paternalistic ties of the agrarian order with the anonymity and insecurity of industrial wage-labour. It substituted the natural rhythms of farmwork and country life with the time and work discipline of the factory system. The interdependent relationships and conventions of the extended family and of the rural community were replaced by the autonomous nuclear family and individual action. In short, individuals and families were left very much to their own devices in a system which had no rules save those of the market place. As Hobsbawm (1968) has put it, 'pre-industrial experience, tradition, wisdom and morality provided no guide for the kind of behaviour which a capitalist economy required' (87). Although these generalisations are taken from analyses of British urbanisation, they are equally applicable to the working-class experience in North American cities. Drawn in huge numbers from the rural and new industrial backgrounds of Europe, and, as time progressed, from the continent's own rural communities, the polyglot of immigrants

11

to the factories and tenements of New York, Baltimore, Philadelphia, Chicago, Montreal and the like faced the same social disruption as their British counterparts (Thernstrom 1973; Ward 1971).

While the move to the city was clearly a socially disruptive process for the working-class masses, it is less easy to determine to what extent this disruption generated nostalgia for the countryside. In the first place, the sentiments of the common people generally are not expressed in recorded form. Secondly, we have to recognise that the business of surviving the social dislocation and poverty of the new environment would have made nostalgia a luxury for most urban immigrants. Moreover, we should be cautious of assuming that for the working classes, urban life was inherently worse than rural. After all, the the old rural order contained its own exploitation and misery, from which the city offered the only escape. On the whole people were better off, at least in material terms, as urban rather than rural workers, especially as the labour needs of agriculture diminished. Nevertheless, as Gans (1982) has shown, working people, especially in the immigrant neighbourhoods of American cities, did attempt in community and familial relationships to re-create some of the elements of their rural origins. Furthermore, as later chapters will reveal in greater detail, they were quick to seek relief from the pressures of urban living in parks and in brief excursions to hike and picnic in the surrounding countryside, especially when public transportation became more widely accessible.

The principal contribution of the working classes to the development of a countryside ideal, however, was not so much through any direct reaction to their own urban and industrial experience as through their relative position in the urban class structure. The formation of a large urban proletariat was inextricably linked with the rise of the middle class; of the entrepreneurs, bankers, lawyers and other professionals who financed and managed the new industrial system which sustained (and, of course, was sustained by) the influx of the masses. It was within this new bourgeoisie, which had already begun to appear as a social and economic force in pre-industrial Britain, that the seeds of reaction to urbanisation and the concomitant sentiment for the countryside were sown.

In common with bourgeois society in general, the middle classes which emerged out of the urbanisation of both Britain and North America were defined by their wealth, status and respectability. Their first concern, therefore, was to express their position by separating themselves, socially and spatially, from the working classes. Initially this took the form of a shift from the usual pre-industrial arrangement of employer and family living on or near the place of business, participating in its daily operation and often absorbing employees into family life, to one which emphasised the separation of residence from workplace. This was partly determined by the production modes of the new industrial system. But it also reflected, in the social climate of late eighteenth-century Britain where it began, the growth of the domesticated nuclear family as the symbol of middle-class morality and status. An ideal which has since become firmly established as the foundation of middle-class values on both sides of the Atlantic, it was in the early nineteenth century strongly influenced by an evangelical movement which promoted the sanctity of Christian family life and the role of women as child-raisers and housekeepers (Fishman 1987). Central to this was the notion of the home as a place of refuge, 'in which the husband could recover from the pressures of business life while wife and children remained inviolate from the temptations of the wicked world' (Burnett 1978: 193). Segregation from this world, from the urban masses and the socially unstable, congested and unhealthy environment which they were perceived to have created, thus became synonymous with the middle-class domestic ideal. In exclusive urban enclaves and, increasingly in suburban villas and country houses the middle classes turned their backs on the industrial city as a residential experience. And, as it expanded as a class through the nineteenth century, this pattern, as we shall see in later chapters was repeated again and again in the reproduction of middle-class values in suburbia and the countryside beyond.

Yet it was not simply the desire for spatial segregation from the industrial city that made the rising middle class such fertile ground for the nurturing of sentiments towards the countryside. This was an increasingly affluent, upwardly mobile, and educated element of urban society. It was thus furnished with the means not only to realise its residential ideal but also to travel and thus cultivate the taste for scenic appreciation and the broader amenities of the non-urban landscape. It could aspire to a country property and so ensure the continued gentrification of the rural landscape. But beyond this, it is important to recognise that the middle classes, by virtue of their growing centrality to the development of industrial society, became, as Raymond Williams (1960) has explained, the main arbiters of cultural values. With their expanding wealth and numbers, together with their strong belief in the importance of a classical education, they became the main patrons of art, literature and music. The Edwardian novelist Arnold Bennett, for instance, thought that the English middle classes 'lacked temperament', but admitted that they bought his books (in Lewis and Maude 1949).

On both sides of the Atlantic, it was the growth of the middle class in the nineteenth century which prompted the extraordinary expansion of what today might be called a cultural industry. The rise of industrial civilisation was accompanied by a huge increase in publishing from books to magazines, by the popularisation of art and music in the public galleries and concert halls which sprang up in cities large and small, and by the invention of new techniques of artistic expression such as photography. With its materialistically driven search for status, it was the middle class which consumed all this culture. Moreover, it was from the middle

13

class that those who created this culture came. While many may have aspired to the avant-garde and the eccentric, the fact is that most of the great literary and artistic figures of the nineteenth century came from and remained firmly in the ranks of the middle class. This general cultural framework was tailor-made for the literary and artistic idealisation of the countryside. It was not only that the middle class was a ready and growing market for culture in general, but that its values were highly receptive to the pastoral imagery and rural nostalgia associated with the search for domestic bliss in tranquil and secure surroundings. It was also consistent with the inherent conservatism of a society in which status and respectability were generally synonymous with a gentrified lifestyle, whether in town or country.

While the rise of middle-class society established the social and cultural conditions within which popular sentiment for the countryside could grow, it also created an intellectual climate from which ideas about country and city, and indeed about the whole basis of modern industrial civilisation could emerge. From the outset, these have been dominated by two complementary strands of thought: the one involving a broad critique of urbanism and industrial progress, the other combining romanticism and agrarianism in a general idealisation of nature and country life.

URBAN HORDES AND DARK SATANIC MILLS

Anti-urbanism and anti-industrialism, or at least a profound ambivalence and distrust of the urban–industrial system is, as a number of scholars have shown in more detail than is possible in this book, a strong and persistent theme running through the intellectual history of western civilisation (Coleman 1973; White and White 1962; Williams 1973). Much of this has been directed at the industrial city itself – at its physical and social conditions, its moral and political shortcomings, its aesthetic failings and its consumptive space-economy. But it has also involved a broader critique of industrialism, modernism and the general ideology of material progress. Both have become pervasive themes in Anglo-American value systems.

From its beginnings the industrial city has been regarded by its harshest critics as a pathological environment; chronically and irrevocably diseased by the very processes of industrial capitalism; 'a pestiferous growth', as William Cobbett called London in 1821, on an otherwise harmonious landscape. Writing from the perspective of his famous tour of an England which he still regarded as an agrarian nation, it is hardly surprising that Cobbett should have described the burgeoning industrial towns that he visited as 'these unnatural embossments; these white swellings, these odious wens, produced by corruption and engendering crime, misery and slavery' (Cobbett 1912: 43). Yet it is this image of the city, of 'the urban inferno, with its hordes of faceless inhabitants', as Steiner (1971: 23) has put it, which preoccupies nineteenth-century observers. Against the liberal and utilitarian proponents of industrial capitalism stood an articulate and expressive body of opinion which stressed the failings of the new urban environment. One of its first and most influential critics, Robert Southey, wrote in 1807 of Birmingham as a city where 'the noise ... is beyond description.... The filth is sickening ... active and moving, a living principle of mischief, which fills the whole atmosphere and penetrates everywhere.' 'It must be confessed,' he continued, 'that human reason has more cause at present for humiliation than for triumph at Birmingham' (in Coleman 1973: 34).

Southey's repeated condemnation of the conditions of the industrial city were echoed again and again during the nineteenth century. The grim picture of the fictional Coketown, portrayed in Charles Dickens's novel, *Hard Times* (1854) with its 'black canal' and 'river that ran purple with ill-smelling dye' (30) became the archetypal image of the Victorian city. In 1866 John Ruskin described London as 'that great foul city ... rattling, growling, smoking, stinking – a ghastly heap of fermenting brickwork, pouring out poison at every pore' (Ruskin 1866: 21). Almost forty years earlier the pessimistic vision of this heap invading the innocent and helpless countryside was caricatured by the cartoonist, Cruikshank (Plate 1.1). And, almost a century later, Lewis Mumford, the great modern critic of urbanisation, was describing nineteenth-century industrialism as producing 'the most degraded urban environment the world has ever seen' (Mumford 1961: 447).

Serious criticism of the Victorian city, however, was sustained not so much by these horrific images of its physical conditions as by a more general concern about its social and political failings. Conservative minds like Southey's saw the poverty and overcrowding of the industrial working classes as a direct threat to the stability of the established social and moral order, and indeed of the urban safety of the middle and upper classes. Radical opinion attacked the city for its exploitation of the working classes and the perpetuation of their appalling living and working conditions. Although both viewpoints were influential in bringing about gradual improvements to urban sanitation and housing, this did little to divert concern over the social degradation of the industrial city. Engels' 1844 study of the English working class, the later novels of Dickens, the writings of Charles Kingsley, even Disraeli's suggestion that mid-Victorian England was divided into two nations of rich and poor, led the way in sustaining the criticism of the quality of urban life in nineteenth century Britain (Coleman 1973). Underlying this was a growing sense, in some intellectual quarters, of the general failure of the city as a human community. The leading proponents of this view were William Morris and Sir Ebenezer Howard. Morris, most notably in his utopian vision of a communal,

15

Plate 1.1   *London going out of Town – or – the March of Bricks and Mortar,*
Thomas Cruikshank, 1827

pre-industrial Britain in his novel *News From Nowhere*, argued for the total abandonment of what he saw as an exploitative and entirely anti-social late Victorian London (Morris 1891). Howard (1898), in his proposals for garden cities believed that the benefits of the industrial city were offset by 'excessive hours of toil, distance from work, and the isolation of crowds' such that the qualities of community were completely lost (197).

It is this concern for the loss of community which has tended to pre-occupy North American criticism of the nineteenth-century city. Early opinions about cities in general, in fact, were strongly influenced by the negative images of the English industrial city. Even before English criticism appeared to any serious degree, Thomas Jefferson was warning against the reproduction of the industrial city in the new republic, a view which retained considerable intellectual support throughout the ante-bellum years (Marx 1964). Indeed, the very development of American cities during this period evinced, in the writings of Melville, Hawthorne and Poe a strong literary fear of the repetition of the human degradation associated with the British Victorian city. And, as American cities did become more industrial-ised, this led to recurrent intellectual complaints about their unpleasant physical conditions (White and White 1962). Yet what Jefferson and other leading republicans feared most in the rise of cities was that their material-ism and corruption made them inimicable to the flourishing of republican morality and democracy. This was founded on the belief that, while small cities and rural districts could be run as orderly, democratic communities, large industrial cities could not (Kristol 1970). A view which has held sway in American political ideology ever since, this has contributed to the retention of small-town idealism in American culture.

Much of this idealism, however, stems also from a growing critique of the social failings of the American city. The soaring rate of post-bellum urbanisation, with its polycultural influx of immigrants and exploitative economic system, generated increasing intellectual scepticism of the sustain-ability of the city as a social organism. It was criticised for its decadence by conservative New England intellectuals like Harvard professor Charles Eliot Norton and writer Henry James (Shi 1985). It was despised by Henry Adams for the power that he saw it giving to immigrant Jews (White and White 1962). By the early twentieth century the problems of the urban community had become the subject of the new field of sociology. Strongly influenced by the ideas of the German sociologist, Friedrich Tonnies (who proposed the extraordinarily durable concept of *gemeinschaft–gesellschaft*, that is of a dichotomy between the communality and homogeneity of small communities and the individualism and cosmopolitanism of metropolitan society), American urban sociologists, notably Robert Park and John Dewey, added academic weight to the belief that the growth of cities had led to the breakdown of community and to general social dislocation (White and White 1962).

Accompanying the critique of the physical and social conditions of the industrial city towards the end of the nineteenth century was a growing condemnation of the geography of metropolitan expansion. Sir Ebenezer Howard's ideas were influenced as much by his concern over the increasingly sprawling pattern of urban growth as by his distaste for the conditions of urban life itself. His proposals for planned decentralisation were complemented by the ideas of Sir Patrick Geddes. A Scottish biologist who had become increasingly disturbed by the rise of the conurbation (he first coined this term), Geddes attracted considerable attention around the turn of the century with his arguments for ecologically-based approaches to civic planning (Geddes 1915). Together, as is revealed in more detail in Chapter 6, the arguments of Howard and Geddes provided the inspiration for a growing campaign within the emerging urban planning profession against the large-scale metropolis. In Britain they formed the basis for a persistent and popular ideology of urban containment. In the inter-war years it was expressed in the rhetoric of Clough Williams-Ellis (1928) who saw urban sprawl as an octopus spreading its tentacles across the countryside, and in the town and country planning ideas of Sir Patrick Abercrombie (1926) (see Chapter 5).

Across the Atlantic, reaction against urban sprawl inspired the formation of the American Planning Association, and in particular the ideas of its leading spokesman, Lewis Mumford. By the 1920s, Mumford had established himself as the principal critic of the American metropolis. Having condemned the industrial city, he directed his criticism to the increasingly sprawling character, the megalopolitan tendencies of American urbanisation (Mumford 1926). Mumford's views were shared by the architect Frank Lloyd Wright, whose utopian and hence unrealised community of Broadacre City represented his belief in the need to decentralise American urban growth. For Wright the modern metropolis was a denial of civilisation (Wright 1958).

It would be erroneous to suggest that the criticisms of urbanisation which have been outlined above were all prompted by purely anti-urban sentiments. Much of it sought improvements to urban living and alternatives to metropolitan forms, rather than the total abandonment of cities. Indeed, as Morton and Lucia White (1962) have observed, the response to urbanism, at least amongst American intellectuals, is divided between the sophisticates and the primitives; between those who wished to reform and perfect the city and those who wished to retreat from it (White and White 1962). Yet, the rhetoric of the critique and the extent to which it reflected the growing middle-class reaction against the urban environment, meant that it was readily incorporated into anti-urban modes of thought and, in particular, into the language of popular culture.

What sustained anti-urban sentiment in Anglo-American culture, however, was that it was underlain by deeper anxieties about industrialism itself; about the very system upon which urbanisation and the kind of cities it produced was founded. With the rise of the industrial system has come a broadening and unremitting debate about its consequences for both the human condition and the state of the earth. One of the central dialectics of modern civilisation, it is a debate which, at its highest philosophical level, sets the ideology of material progress against that of nature and human spirituality. The former emanates from the rationalism of Thomas Locke and the empiricism of David Hume which emerged in late eighteenth-century Britain as the underlying philosophies of industrialism. Hume's belief in the liberating value of materialism provided the foundation for the rise of Benthamite utilitarianism in the early nineteenth century, which, in general terms, argued for the utility of economic progress as the means of securing the happiness of the greatest number. These ideas, together with the political economies of David Ricardo and Adam Smith, generated considerable support amongst those who foresaw the benefits of British industrial supremacy. One of the most outspoken defenders of industrialism was Thomas Macaulay, a Whig politician who in the 1830s argued that 'people live longer because they are better fed, better lodged, better clothed, and better attended in sickness; and that these improvements are owing to that increase in national wealth which the manufacturing system has produced' (quoted in Coleman 1973: 63). The philosophy of progress acquired an immediate following in early Republican America. Led by Alexander Hamilton, who was a keen disciple of Hume, capitalist expansionists saw individual economic success and the factory system as the basis of social progress and national power (Shi 1985). The machine, as Leo Marx (1964) has stressed, became in the early decades of the nineteenth century a symbol of American progress, a means of enriching the mind as well as the body, and of realising the national ideal. In Britain, too, the belief in the inevitability of human progress based upon the technology of industrialism, and celebrated in Paxton's glass cathedral at the Great Exhibition of 1851, became part of the faith of the Victorian age.

The arguments against industrialism orginate in the Whig and Radical dissension of the late eighteenth and early nineteenth century. Against the philosophy of Hume were set the protests of Thomas Paine and William Cobbett: the former eventually locked up for his condemnation of the new economic system's perpetuation of the old social and political order, the latter achieving notoriety for his protests against industry's exploitation of the agrarian economy and its displacement of rural society. And against the boosterism of Alexander Hamilton was set the conservative republicanism of John Adams and Thomas Jefferson, who opposed the greed, material excess and, above all, selfishness which they saw as the inevitable consequences of industrialism (Shi 1985). Jefferson was not opposed to technology as such, indeed he saw great benefits in its application to the development of the American economy. What he objected to, in a philosophy which has remained remarkably durable in American ideas ever since, were the corrupting effects of the industrial system as it was developing in Europe on the moral fibre and civic spirit of society (Marx 1964).

As industrial capitalism tightened its grip on economy and society, intellectual doubts about its consequences appear increasingly to have overshadowed arguments in its favour. By the 1830s Thomas Carlyle was advancing his influential ideas on the oppressive effects of industrialism on the human spirit, a notion which in 1844 was taken one step further by Karl Marx's treatise on the alienation and devaluation of the human experience by materialism. Carlyle's ideas attracted the attention of the American transcendentalist Ralph Waldo Emerson, who actually visited Carlyle in England in 1833 (Marx 1964). Although he saw its potential for improving the human condition (and indeed wrote in celebratory terms of the marvels of modern machines), Emerson maintained that technology should never subordinate the pursuit of self-fulfilment (Shi 1985). Emerson's ambivalence about industrialism, however, was transformed into full-scale rejection by his protégé, Henry Thoreau. Thoreau's celebrated retreat to the Concord woods symbolised his outright repudiation of the culture of technology, in which 'man has no time to be anything but a machine', and in which the pursuit of material comfort and wealth were 'positive hindrances to the elevation of mankind' (Thoreau 1854: 9).

It was in the late nineteenth century, however, that the ideological rejection of industrialism reached its zenith. Leading the way in this was John Ruskin, whose hostility to *laisser-faire* economics and the materialism and ugliness of modern technology initiated an anti-industrial, anti-modernist movement in Britain. To Ruskin's conservative philosophy was added William Morris's socialist condemnation of industrial capitalism. With Ruskin, he prompted the rise of the arts and crafts movement which sought to replace the impersonal and aesthetically sterile products of mass production with those of craftsmanship and good taste. 'The leading passion of my life,' wrote Morris, 'is a hatred of modern civilisation', a sentiment which he cast in the sharp idealistic relief of *News From Nowhere* (Morris 1891). Morris was not alone in criticising modern industrialism from a utopian perspective. Three years before *News From Nowhere* the American Edward Bellamy had published *Looking Backward*, a futuristic vision of a perfect metropolis which reflected his belief that his own society was doomed (Bellamy 1888). In deeper political vein, there was the socialist utopianism of Leo Tolstoy and Peter Kropotkin, who from somewhat different perspectives argued for communitarian alternatives to industrial capitalism.

Martin Wiener (1981) has advanced the thesis that English culture reflects an inherent distrust of the whole notion of industrial progress. Presented with the choice of being the 'Workshop of the World' or a 'Green and

Pleasant Land', educated society in late Victorian and Edwardian England, he suggests, increasingly chose to identify with the latter. He traces this, in part, to the radical anti-industrialism which has been described above. From Ruskin and Morris, he observes, there develops a tradition of conservative, liberal and socialist proscription of Victorian industrial capitalism which persists well into this century. It is reflected in the Edwardian writings of intellectuals such as G. M. Trevelyan, Hilaire Belloc, C. F. G. Masterman and G. K. Chesterton who, in Masterman's words, saw the nation's social standards being eroded by the 'new dissolvents of the nineteenth century: urbanism, industrialism, and cosmopolitanism' (quoted in Wiener 1981: 60). Socialist reaction was led by the Fabian historians, Beatrice and Sidney Webb who saw in the rise of industrialism the exploitation of the working classes (Webb and Webb 1923).

Wiener, however, argues that the decline of the English industrial spirit should be attributed primarily to the gentrification of the middle class; to its repeated absorption into the conservative value system of the aristocracy and the landed gentry. These values were sustained by the social dominance of provincial life and by public school traditions which prepared its students more for careers in the military, the diplomatic service, the higher professions and the Church than in science or technology. They were fostered by an intellectual and literary tradition which either ignored the industrial foundations of English power or rejected it altogether. Wiener's central thesis is that the English, at least until 1945, regarded industrialism and modernism as inherently un-English. The real heart of England, as Prime Minister Stanley Baldwin observed on numerous occasions, lay in the traditions and landscapes of the countryside. This thesis is supported by Ecksteins' argument that while Germany saw the Great War as a crusade for modernist ideology, Britain fought it to defend the traditions of the old pre-industrial order (Ecksteins 1990).

This theme continues into the Second World War. While 'Nazi Germany was being portrayed as an industrial society run amok, England was seen as the opposite; humanely old-fashioned and essentially rural' (Wiener 1981: 77). Perhaps the most vivid example of this are the posters, one of which is shown in Plate 1.2, which were designed to bolster the morale of troops abroad and which represent the ultimate denial of the reality of the Britain from which most of those troops came.

Although the rise of industrialism in North America clearly produced misgivings about its impact on society, it certainly did not translate into the anti-modernist sentiment which Wiener has identified in Britain. At the turn of the century the USA was emerging as the world's greatest industrial power. Nowhere was the culture of technology and modernisation more enthusiastically embraced. Nevertheless much of the anti-urbanism which surfaced with the likes of Mumford and Lloyd Wright reflected a growing sense of the misapplication of science and technology. And, as North

Plate 1.2   A Country Worth Fighting For: British Second World War poster

America descended into the Great Depression in the 1930s, doubts about the industrial system itself began to be expressed more explicitly. Coleman has observed that anti-urban sentiment in nineteenth-century Britain seems to have followed a cyclical pattern, rising and falling with the fortunes of the economy (Coleman 1973). This thesis can be applied equally to twentieth-century responses to industrialism, not just in North America but in modern western thought in general. And, as this century has stumbled from one devastating attack on civilisation to another, scepticism of the technological order has become a recurrent ideological theme. To the nineteenth-century attacks on the alienation of industrialism have been added an increasingly radical repudiation of the whole fabric of industrial capitalism (Shi 1985). It includes the philosophical critique of technocracy in the writings of Jacques Ellul, George Grant and William Irwin Thompson. Within the realms of social analysis it has developed as an attack on the values of consumer society in the arguments of Daniel Bell, Theodore Rozcak, Kenneth Boulding and Christopher Lasch. Most completely as a denunciation of western civilisation, there have arisen the prophets of the new environmentalism, such as Aldo Leopold, Barry Commoner and E. F. Schumacher. And, finally, it has remained etched in the public mind in the frightening vision of the technocratic dystopias of George Orwell's *1984* and Aldous Huxley's *Brave New World*.

NATURE WORSHIP

Nature, or more accurately, the objects of nature which make up a pleasing natural scene, has long been the counterpoint of reaction against the city. As Tuan (1974) has argued, the dichotomy between the natural and the artificial, between the works of God and those of man stretches back to the earliest recorded civilisations. In ancient Greek and Roman culture, and of course in the creationist foundations of the Judeo-Christian tradition, this was largely expressed in metaphorical terms. Yet classical pastoralism, as we have seen, was also influenced by the actual experience of the sights and sounds of nature. From this imaginary and experiential mix, nature emerged as the defining focus of the theme of retreat from the civilised world. This reappears in the rise of post-Medieval urban society; in the revival of literary pastoralism and in a growing attraction to nature as a source of pleasure. For the most part, however, it was the objects of nature – birds, trees, flowers, streams – which were the main source of attraction. The anthropocentric Renaissance mind, however, had a hierarchical view of the world which placed nature in an inferior position to that of human civilisation. There was little interest in nature for its own sake; wild nature, in particular, evinced fear and loathing. As a source of pleasure, therefore, nature was placed in the detached and selective setting of the garden and the farm. This attitude achieved its fullest expression in the formalism of the Italianate garden, and persisted in the contrived settings of the Dutch landscape school of painting in the sixteenth and seventeenth centuries.

Changing attitudes to nature

With the eighteenth century came fundamental shifts in attitudes to nature. In the words of Clarence Glacken (1967), 'In no other preceding age had thinkers discussed questions of nature and environment with such thoroughness and penetration' (551). That this roughly paralleled the changes which led to the rise of industrialism is no coincidence. In the first place it was a product of the very intellectual turmoil from which the scientific underpinnings of the industrial revolution emerged. The Enlightenment was the age of both invention and ideas. A significant body of these ideas sought to understand how, with its new potential, human civilisation related to the natural environment. Secondly, the very development of industrialism itself gave the human experience of nature new symbolic and sensory meanings.

Glacken's great study of nature and culture in western thought provides arguably the most coherent interpretation of the confusion of ideas about nature which arose in the eighteenth century (Glacken 1967). What follows, therefore, is drawn mainly from Glacken's book. Essentially, the eighteenth century saw a philosophical reappraisal of the teleological and Cartesian approaches to nature. In place of the teleological belief that nature existed to fulfil God's grand design for a human-dominated earth, and of the Cartesian notion that this design could be understood as a geometric and mechanical order, came the idea that nature was governed by its own forces of which the human species was a part. This idea had, in fact, already been proposed in the mid-seventeenth century. It was with the work of the Swedish botanist, Linnaeus, in the early eighteenth century that this idea gained currency. Linnaeus anticipated the science of modern ecology by seeing the earth as a self-renewing and self-cleansing system. The most direct consequence of this was the development of a more scientific approach to the study of nature based on improved classifications of plant and animal species. Linnaeus's famous botanical scheme encouraged a growing interest during the eighteenth century in the identification and recording of species. Natural history became both a popular pastime and a serious search for a more accurate understanding of the workings of nature.

The most profound shift in ideas about nature, however, must be attributed to the philosophies of David Hume and Immanuel Kant. Hume's views on the 'economy of nature', with its unsentimental internal efficiency, and Kant's belief in human causality with nature (he also argued that nature could exist quite happily without humans) represented the most complete rejection of the teleological argument. These ideas were given added momentum by Rousseau, whose notions of the determining role of the natural environment on the perfectability of the human condition brought nature into the forefront of civilisation. Essentially what these three great minds understood was the interdependency of nature and humans. By the end of the eighteenth century the old view of nature was rapidly being eclipsed by the new. The new ideas were given added credence by the phenomenal increase in the global knowledge of nature that had come with the rise of natural history. The work of botanical explorer like Sir Joseph Banks (the botanist on Cook's voyages) and Alexander von Humboldt served to increase the understanding of nature as an ecosystem. To this must be added the work of Thomas Malthus who in his profoundly influential essays on population, published in 1798, rejected the idea of the plenitude of nature. Malthus was the first to recognise the natural limits of the earth and the holistic basis of its natural systems. Above all he raised the most serious doubts about human ability to manipulate the natural environment to its own material ends.

By the time Darwin had expounded his theories of natural adaptation fifty years later, the concept of a natural environment governed by its own internal order and imposing limits on the progress of human civilisation was reigned directly against the belief in the unlimited power of science and technology. Anthropomorphic values remained but, at the same time nature came to exist in its own right. Already by the end of the eighteenth century the humane treatment of animals and the protection of plant-life had become an important issue (Thomas 1983). By the early nineteenth century the study of natural history had reached new heights of popularity and had become increasingly preservationist in its tone as the human impact on the environment became more fully understood. Herein, of course, lie the philosophical roots of the modern environmental movement with its belief in the restoration of the natural order as the basis for remedying the ills of industrialism.

Accompanying the shift towards greater ecological understanding in the eighteenth century was a growing awareness of nature's significance for the human experience. The observation and description of nature took on a tone of delight and awe, from which emerged the notion of its sublime powers. And, as the mechanism of industrialism expanded, the search for human identity and fulfilment became increasingly centred in the natural world. The sublime view of nature first appears at the beginning of the century in the writings of the scholarly Earl of Shaftesbury. For Shaftesbury nature was a vision, 'whose looks are so becoming and of such infinite grace; whose study brings such wisdom and whose contemplation such delight' (in Thacker 1983: 16). As a deist who believed that it revealed God's perfection, Shaftesbury elevated nature to divine status. What is more important, however, is that he replaced the objective portrayal of nature, which dominated neo-classical thinking, with subjective feeling, with, that is, the sensibilities as a means of perception (Furst 1971). Nature thus acquired the qualities of the sublime – lofty, mysterious, a state of experience rather than of order and logic (Frye 1967).

The idea of seeing sublime properties in nature was immediately popularised by commentators such as Joseph Addison in his columns in the *Tatler* and the *Spectator*, and, as the next chapter shows in more detail, in the work of poets and artists. What all this led to was a wilderness cult, for it was only in its wild state, as Burke argued in his famous intellectualisation of the relationship between the beautiful and the sublime, that the beauty of nature could stimulate the emotions (Hussey 1967). Stirred by poetic language and artistic images rather than by Burke's erudition, affluent society began to seek out the inspiration of wild scenery, first in the awesome grandeur of the Alps and then increasingly in the gentler 'wildness' of the Lake District and the Scottish Highlands (Thomas 1983). From this emerged the idea of the picturesque, in which the notions of the beautiful and the sublime became merged into a aesthetic conception of what constituted a perfect natural scene.

Romanticism and transcendentalism

With the second half of the eighteenth century came a philosophical and artistic shift in values which, instead of seeing humans as social beings gathering inspiration from the divinity of nature, sought to find human identity and consciousness within nature (Frye 1967). It is this shift which, in very general terms, underlies the rise of romanticism. Its philosophical roots are in Diderot and especially Rousseau. For Diderot, creative genius drew inspiration from nature rather than society. Rousseau took this idea further by arguing that general human fulfilment was dependent on living a life which was natural. He believed that humans should first seek identity with nature, live in the 'natural state'. Only then could the perfect human society be formed. Rousseau's belief in the importance of returning to nature was influenced by his growing antagonism towards eighteenth-century civilisation. It was, of course, a reaction against the forces of rational science and economics; the guiding principles of industrialism. And this is the determining context of much of the romantic attitude towards nature which came into full bloom at the end of the eighteenth century.

It is expressed primarily in the work of the English romantic poets: Coleridge, Clare, Crabbe, Blake and, above all, Wordsworth. It is Wordsworth, of course, who with Blake became the icon of English romanticism. But while Blake was busy exploring nature's darker side, it was Wordsworth who revealed what Northrop Frye (1967) has described as the 'moments of feeling' and the 'expanded consciousness', the gnosis that comes from the penetration of nature's 'huge and mighty forms'. Wordsworth together with his contemporaries Coleridge and Southey, saw little to praise in the social conditions of industrial society. Increasingly melancholy about his own ability to come to terms with this, he withdrew into the solitude of his beloved Lake District. For Wordsworth nature was the ground of life and the symbol of eternity. But, above all, it was the restorative for all who had suffered a breakdown of faith in themselves and society.

For a growing middle class attempting to come to terms with the social and economic turmoil of the times, the romantic view of nature had a distinct appeal. The popularity of the Lake Poets had much to do with the insatiable demand for new cultural experiences. But it was also a product of the moral conservatism of a class which was more than ready to turn to nature as a way of escaping responsibility for the problems of industrialism.

By Victorian times, the romanticisation of nature had reached new heights, fuelled as it had been for several decades by a rising tide of poetry, art and music which drew upon nature for its creative inspiration. In the New World, the seemingly limitless expanse of wilderness, of pure nature, presented the most obvious possibilities of applying the inspirational ideals of romanticism. Alongside the desire to tame and exploit nature, there arose in the early nineteenth century a growing reverence for the American wilderness which drew directly upon the European romantic tradition and especially on the work of the Lake Poets (Nash 1967). William Cullen Bryant, arguably the founder of American romanticism, read Wordsworth for the first time in 1817 and immediately 'a thousand springs seemed to gush up at once in his heart' (Huth 1957: 31). With Bryant's celebrations of the grandeur of the Catskills and the Pallisades, James Fenimore Cooper's stories of the noble backwoodsman living in harmony with the forest, and the picturesque painter Thomas Cole's scenes of untouched wilderness, nature worship became a popular pastime. Cole, too, was a disciple of Wordsworth, becoming aware in his search for picturesque scenery, of the 'sublimity of untamed wilderness and the majesty of eternal mountains' (Huth 1957: 50).

The romantic approach to nature in American thought, however, reaches its peak with the rise of transcendentalist philosophy. Drawing to some extent on Platonic idealism and eastern mysticism, but more directly on the Wordsworthian belief in the spirituality of nature, the American transcendentalists argued that by re-establishing contact with nature, and especially wilderness, humankind could transcend physical existence and discover the perfection of the inner being. Although it included other figures, notably George Ripley, who, in his Brook Farm community attempted to put transcendentalist principles into practice, the dominant protagonists of this philosophy of nature were Emerson and Thoreau. Emerson first expounded his views in a 1836 Boston lecture entitled 'Nature' and over the next thirty years established himself as the chief expositer of transcendentalism and the prime intellectual force behind American attitudes to nature. Central to Emerson's philosophy was the belief in the mystical union of God through contact with nature; that 'behind nature, throughout nature, spirit is present' and that 'the Supreme Being does not build up nature around us but puts it forth through us' (in Huth 1957: 88). In other words, in nature lay the true salvation for the human soul.

While Emerson was the intellectual force behind transcendentalism, his protégé, Henry David Thoreau, was its main practitioner (Koster 1975). In his famous retreat to the cabin by Walden Pond, Thoreau placed the idea of nature as the source of spiritual fulfilment in the context of personal experience. Motivated, as we have seen, primarily by his despair over the enslavement of human society by materialism, Thoreau believed that true freedom could be achieved only through self-reliance, simplicity and living in close contact with nature. These were attractive ideas to a mid-nineteenth-century society trying to make sense of a rapidly changing world. Although he wrote several other pieces in which he expounded his enthusiasm for the inspirational properties of wilderness, it is with *Walden* (1854) that Thoreau's ideas have become most readily identified. With this little book, which in scores of reprintings and millions of copies over the years has become the bible of nature worship, Thoreau immediately attracted a following from a widening circle of readers (Nash 1967). Included among these were John Burroughs, whose nature essays did much to translate Thoreau's ideas into a popular enthusiasm for the study and appreciation of nature, and John Muir, the inspiration behind the early national parks movement, about whom more is written in Chapter 6.

With the profound shifts in ideas which occurred during the eighteenth and nineteenth centuries nature was elevated to a sacred status (Tuan 1974). With new ecological understanding and new perceptions of its inspirational powers, the natural world came to be viewed as the salvation for the alienating conditions of industrialism. However, while the philosophy behind this undoubtedly filtered down into the value system of educated society, much of the popular attraction to nature stemmed from its aesthetic appeal rather than from its ideological associations. While poets and philosophers were arguing its virtues for the inner soul, it was nature's contribution to the creation of pleasant scenery which increasingly enthused the public at large. In the eighteenth and early nineteenth centuries the growing interest in wilderness was sustained more by the values of the picturesque than by the higher principles of the sublime. When people began to flock to the Alps and the Lakes they did so in search of the scenery which they associated with the landscapes of Claude, Poussin and Salvator Rosa, possibly viewing it through a so-called Claude glass which permitted the arrangement of the scene as if in the frame of a painting. It was the picturesque qualities of the wilderness, of the natural scenes of the Catskills and Adirondacks, expressed most notably in Thomas Cole's paintings, which also stimulated American enthusiasm for nature travel (Huth 1957). Furthermore, the notion of the picturesque did not remain with wild scenery. It increasingly became applied to any landscape which presented an attractive scene, and in particular to what, in the context of early nineteenth-century America, Leo Marx has termed the 'middle landscape'; the farmscape between the artificial city and the wilderness (Marx 1964). While Marx failed to recognise the growing strength of American attachment to wild scenery during the nineteenth century, he did point to the popular, sentimental side of romanticism in which natural and pastoral scenery become synonymous.

The dilution of picturesque and romantic interpretions of nature into popular landscape taste is nowhere more apparent than in England. By the beginning of the nineteenth century the term picturesque had come to apply to settled as much as to wild landscapes. Indeed, the very absence of truly wild landscapes ensured that it would be nature in domesticated settings with which the English would most readily identify. This is reflected in the application of the conventions of the picturesque to the landscaping of country estates in which informal planting and careful arrangement of vistas was intended to achieve the effect of a natural scene.

It is reflected also in a shift in romantic art and literature to the celebration of nature in pastoral scenes.

AGRARIAN SIMPLICITY AND RURAL NOSTALGIA

The absorption of nature worship into a more general sentiment for rural scenery can be attributed in large measure to the belief that rural life is more natural than urban; that, by virtue of their closeness to the soil and their dependency on the physical environment, farming folk live a more natural and therefore more fulfilled existence. This idea, as we have seen, is an important element of romantic philosophy stemming first from Rousseau's ideas about natural community and culminating in the Thoreauvian belief in the naturalness of simple living. At heart it is an idea which romanticises pre-industrial culture, casting the traditional rural lifestyle and communities of the past in nostalgic contrast to the dynamic and individualistic culture of the present. Perhaps the most important element of this is the attachment of reverential status to farming as a way of life. As we saw at the beginning of the chapter, this is an ancient belief which explains much of the anxiety about the rise of urban society. It has also been a recurrent theme in modern western culture. While it has influenced the British countryside ideal, it is on North American attitudes to the countryside that it has had the most significant impact.

Most migrants to the New World in the seventeenth and eighteenth centuries were seeking new beginnings; escapists from religious intolerance and political repression, idealists and adventurers looking for new ways of living and opportunities to make it rich, above all people seeking freedom from the social hierarchies of Europe. These values had to be adapted to a quite different geographical setting and accommodated within the broader mercantile and political objectives of the colonising powers. The pattern of rural settlement and land use which emerged from this emphasised the virtues of individual freedom. Certainly, initial European colonisation attempted to re-create the communalism of peasant society. The first generation of Puritan settlements in New England was organised along village lines (Meinig 1986). Against these centralising tendencies ran the forces of individualism which, in the context of an apparently limitless supply of land, led to an increasingly scattered pattern of farm settlement (Wood 1982). Attempts to organise settlers into village-based communities in other areas, such as Pennsylvania and New France (Québec), were also thwarted by a general desire for freedom from the institutional restrictions of village life (Harris 1966; Lemon 1972).

Although these independent instincts were related in part to the rise of religious dissent, they also reflected, at least in the American colonies, the values of migrants who had come from a British rural society in which feudal ties were already breaking down in favour of the individual family farm. In North America individual and family autonomy could achieve its full expression. 'Everyone strove to get out on the land, his own land' (Lemon 1984: 94). The fulfilment, indeed very often the stimulation of these desires was made possible by the supply of cheap and plentiful land on the new continent. In Europe it was labour that was cheap and land scarce, a situation which guaranteed a hierarchical rural society dominated by a land-owning élite. In North America, labour was scarce and land cheap (and often free). With great tracts of land for the taking, mainly from an unconsidered indigenous population, its ownership was available to all who were willing to work it. Moreover government policies ensured relative ease and equality of access to land, especially as settlement spread westwards (Meinig 1986).

From this settlement ethic emerged a society dominated by the individual family farm. The only exception was in the southern plantation system. Elsewhere, the initial clearance of land, the establishment of cultivation, the development of commercial agriculture and with it the formation of strong local economies, were carried out by individual families on their own separate land parcels. This pattern was repeated again and again as new territories were occupied. Until late in the nineteenth century most Americans and Canadians lived either on farms or in the small towns which served them. The family farm ideal is therefore deeply etched in the North American psyche. It is an ideal which, at least in the USA, has been fostered by a powerful agrarian ideology which goes to the very heart of American democracy.

The vision of America as a natural paradise which would be made bountiful by the efforts of the virtuous husbandman dominated eighteenth-century thought. By the time of the revolution it had become one of the founding myths of Republicanism. Puritan revivalists like James Warren of Plymouth argued that 'public virtue . . . must be looked for in the sober and manly retreats of husbandmen and shepherds, where frivolous manners, commerce and high stages of civilisation, have not debauched the principles and reason of mankind' (in Shi 1985: 76). De Crevecouer's *Letters from an American Farmer* celebrated the simple pleasures of the independent farmer drawing his exuberancy from the soil. Benjamin Franklin, influenced by the French physiocrats, argued that agriculture must be the basis of the American economy. But of course it is with Thomas Jefferson that agrarian ideology is most directly associated. In declaring, in his much-quoted phrase, that 'those who labour the earth are the chosen people of God', Jefferson saw the independent yeoman farmer as the source of moral and civic responsibility; the backbone of democracy. The farmer would also be the backbone of a Republican economy which would have no need of manufacturing (Marx 1964).

In the early part of the nineteenth century agrarianism was a term used to describe populist farmers' movements campaigning for the subversion of the prevailing landownership structure. Even Jefferson had no sympathy for these 'agrarian and plundering enterprises'. Jefferson's own vision of an harmonious agrarian republic populated by frugal and hard-working yeoman farmers, however, came to dominate American values throughout the nineteenth century, and indeed has remained firmly fixed in the minds of Americans ever since (Rohrer and Douglas 1969; Wheeler 1976). It has, however, followed several different strands. There is firstly what has been termed a progressive agrarianism associated with the increasingly commercialised character of American farming (Hofstadter 1966). Jefferson himself came to recognise that a productive, commercial agriculture was essential to the security of the republic. And, with the westward spread of settlement, an ever-expanding agricultural society, founded on the commercial success of the independent farmer became firmly entrenched in both the economic and political ideology of the nation (Nash 1967). Even today the United States Department of Agriculture, despite policies which support agribusiness, perpetuates the myth that the family farm is the backbone of the agricultural economy.

More influential on popular thought, however, is the fundamentalist version of agrarianism; the ideal of the simple yeoman farmer as the standard bearer of good Christian, American, democratic values. It has also been reinforced by a pioneer myth which, from the early Puritan belief in the virtues of the hard-working and frugal settler carving a living out of the wilderness, to the Turnerian thesis of the frontier as the defining force of American identity, has established the pioneer farmer as the true founder of the nation. Between the 1880s and the 1930s the fundamentalist viewpoint, in a revival of the Jeffersonian ideal, attracted widespread support. It was promoted by populist farmers' movements like the Farmers' Alliance and the Grange. It was brought to the intellectual forefront by a group of southern agrarians who in a 1930 book entitled *I'll Take My Stand: The South and the Agrarian Ideal* argued against Yankee industrial domination and in favour of a southern way of life based upon the spiritual fellowship of farm life (Lora 1971). Fundamental agrarianism became one of the clarion calls of the New Deal with Franklin Roosevelt declaring, 'In all our plans we are guided . . . by the fundamental belief that the American farmer living on his own land, remains our ideal . . . the source from which the reservoirs of the nation's strength are constantly renewed' (in Shi 1985). Politicians and in particular presidential candidates, have in fact traditionally referred to their farm and rural roots as evidence of their honesty and their down-to-earth character. Finally, the belief in the fundamental virtue of farm life has been sustained by a recurrent back-to-the-land movement. As Chapter 3 shows in greater detail, there has been a procession of advocates for returning to the life of honest husbandry.

Although, as Chapter 3 also shows, the American back-to-the-land movement was paralleled by a similar movement in Britain around the turn of the century, British culture has never embraced either progressive or fundamentalist agrarian ideology to the same degree as American culture. Indeed for the most part British sentiment for the countryside has tended to turn its back on its working elements, and in particular on the idea of farming as an economic enterprise or the foundation of the nation's values. Yet what the British and North American countryside ideal do have in common is a nostalgia for a sentimentalised version of farm life. It is this version of the agrarian ideal, rather than its more ideological counterpart, which has tended to inspire both the actual and the imaginative retreat from city life. For a long time, for example, it was the basis of the British belief that the farmer was the kindly guardian of the countryside, until, of course, it became apparent to post-war preservationists that farming was an industry which could intrude on a pastoral scene as effectively as urbanisation (Shoard 1985). In North America the image of the idyllic quality of farm life lies at the heart of the rising public outcry over the urbanisation of farmland and farm bankruptcies. And on both sides of the Atlantic, it has long persuaded some of the more romantically-inclined citizenry to move to a parcel of land on which they can live out the fantasy of farm life for a few hours each weekend, or to abandon city life altogether in favour of scratching out a living on a farm which, in many cases, its previous owner has been only too happy to sell.

The sentimentalisation of farm life is part of a broader nostalgia for traditional rural life in general. At an intellectual level this has involved a recurrent call for the revival of the old, pre-industrial ways of living. The rise of American agrarian ideology, as we have seen, was closely associated with the Puritan belief in the virtues of simple rural life as against the greed and cosmopolitanism of the city. The transcendentalism of Emerson and Thoreau was a call for living a life of simple subsistence in harmony with nature in a middle landscape between the city and the wilderness (Marx 1964). The anti-urbanism of the New England intellectuals later in the century was based on nostalgia for an earlier America when, in the words of their leading spokesman, Charles Eliot Norton, 'habits of life were simpler; the interests of men less mixed and varied: there were more common sympathies, more common and more controlling traditions and associations' (in Shi 1985: 168).

These sentiments had also come to the fore in the growing anti-industrial ideology of late nineteenth-century Britain. Ruskin and Morris looked back to what they saw as the harmonious peasant society of Medieval England for their alternative to the ills of Victorian culture. These ideas were not new for they drew on a long tradition of reverence for traditional rural life stretching back to the laments for the passing of the pre-enclosure village community in Oliver Goldsmith's *Deserted Village* (1770) and then in William Cobbett's famous *Rural Rides* in 1830. They were also part of a radical tradition of communally-based rural utopianism. Between them,

Ruskin and Morris, together with other radical thinkers, notably Edward Carpenter and Robert Blatchford, were responsible for the broadening intellectual nostalgia for old rural England (Wiener 1981). William Morris probably did more than anybody to transform this nostalgia into a national cult. 'Old English' quaintness and rusticity became the dominant style in everything from architecture to furnishings; indeed there was a veritable William Morris craze in interior design. The English country cottage became a fashionable style both in suburbia and in the trend for rustic country retreats. The arts and crafts movement also inspired a growing interest in the revival of country crafts and traditions, from basket-weaving and cooking to folksongs and dances (Wiener 1981). Furthermore its influence spread across the Atlantic where it spawned several rural arts and crafts colonies, and stimulated much of the early twentieth-century interest in country living.

Back in its country of origin, however, the arts and crafts movement became part of a broader rise of nostalgia for old rural England. The anti-industrial spirit which, Wiener has argued, had come to dominate English middle-class values by the beginning of the First World War, was founded on a vision of a serene rural England. It was a vision promulgated not only by historians convinced that the heart of the nation lay in its pre-industrial past, but also fostered, as the next chapter will show, by a proliferation of books and articles celebrating the traditional character of the countryside. It also became part of the political rhetoric of the nation, most notably in the person of Prime Minister Stanley Baldwin for whom the true England in 1926 was 'the tinkle of the hammer on the anvil in the country smithy, the corncrake on a dewy morning, the sound of the scythe against the whetstone, and the sight of the plough team coming over the brow of the hill . . . for centuries the one eternal sight of England' (Baldwin 1926: 7).

Nostalgia for traditional rural life continued to thrive right up to the Second World War, in the books of country writers and in the growing preservationist movement. And since the war it appears to have gained in strength, both in the proliferation of new books describing the old country ways and reprints of pre-war country nostalgia, and in the phenomenal rise of interest in reviving and disseminating traditional methods of farming, rural crafts and country ways. A similar trend is apparent in North America. At its most serious level it represents the reinvocation, in the context of the whole earth ideology of the new environmentalism, of Jefferson, Thoreau and Morris all rolled into one alternative lifestyle philosophy, with the return to simple, traditional country life as its principal mode of expression. At a more popular level, it is manifested in the extraordinary growth in the preservation of rural heritage in agricultural museums, restored villages, craft centres, country parks and so on. And finally at its most trivial level it has become a consumer industry in its own right in everything from country cookbooks to the reproduction of country living in the gentrified rural communities of exurbia.

IDEALISED COUNTRYSIDES

The main purpose of this chapter has been to explore the historical, ideological and cultural foundations of the countryside ideal. The question that remains, however, is what precisely is this ideal? As we have seen, it has emerged as a product of and a reaction against the rise of the urban–industrial system. It is an ideal, therefore, which is based both on abstract values and on real images. On the one hand the countryside has become a symbol of the good side of civilisation; the locus of human fulfilment and true community, of harmony between nature and humankind, of the virtue of simpler epochs. On the other, it has developed as a set of cultural and landscape images drawn in sharp contrast to those of the industrial city and woven into the conversion of the countryside ideal into actual experience. The countryside ideal is therefore a complex mix of myth and reality, encompassing at one end of the spectrum profound philosophical questions about modern civilisation and at the other, simple escapism.

We must also recognise, however, that the ways in which the countryside has come to be idealised are directly related to the processes and values which have forged its physical and cultural landscape. This is clearly illustrated in the differences between the countryside ideals of Britain and North America. While, as we have seen, they are tied together by similar philosophical threads, and while there are strong parallels in the manner of their cultural expression, the British and North American ideals are also founded on somewhat different perceptions of their respective countrysides.

The British, and especially the English countryside is valued primarily as a landscape aesthetic. This is evident in virtually every aspect of its idealisation from its literary and artistic treatment to its use as an amenity and the campaign for its protection from development. In part this is attributable to the inherent beauty of a compact and thoroughly domesticated landscape of astonishing local diversity and historical depth. However, the aesthetic appreciation of the countryside is inseparable from the social order which created it. As we have already seen the bulk of the English countryside as we know it today was created by the process of enclosure and gentrification which accompanied the spread of landed estates. Although it evolved within the framework of agricultural progress, it was a form of progress constrained by the entrenched hierarchical structure of rural society, in which agrarian objectives were often subordinate to the requirements of gentrification.

From this emerged a countryside in which the preservation of the status quo, and the creation of a landscape for leisure and aesthetic enjoyment became as important as agricultural productivity and rural trade. This established the perfect conditions for the cultivation, literally and figuratively, of the picturesque character of the countryside. As the working population left rural areas in growing numbers during the nineteenth and early twentieth century, the countryside became even more amenable to being appreciated for its scenic quality. Even the rising nostalgia for traditional rural life, for 'olde' England, was largely absorbed into a landscape rather than a social ideal. When Stanley Baldwin celebrated the old country ways he did so in terms of an idyllic scene, of the sights and sounds of the smithy and the plough team. And, when a few years later Sir Patrick Abercrombie formed the Council for the Protection of Rural England it was the appearance of the countryside which he was most concerned about protecting. This then is the archetypal English countryside ideal; a taste, as Lowenthal and Prince (1965) have put it, for 'landscapes compartmented into small scenes furnished with belfried church towers, half-timbered thatched cottages, rutted lanes, rookeried elms, lich gates, and stiles' (87); a timeless landscape in which social and economic realities have been painted out by a picturesque brush.

As I suggested in the Introduction, the word countryside has not carried with it the same emotional connotations in North America as it has in Britain. Indeed it is fair to say that in both the USA and Canada there is a poorly developed sense of the settled countryside as a landscape ideal. To some extent this is a function of the geography of the continent; of its vastness of scale, relative topographical uniformity across large regions and even emptiness, compared with the compactness, local diversity and complete domestication of the British Isles. It can be traced, too, to the youthfulness of its cultural landscape. Yet, like Britain it is strongly associated with the nature of the settlement process itself. Forged out of the edicts of pioneering and survival and developed predominantly around the largely egalitarian and individualistic society and economy of the family farm, the North American countryside evolved as a utilitarian landscape of work and commerce. While the British countryside was being depopulated in the nineteenth century, rural North America was still expanding as farm settlement spread westwards. Settlement followed the rectilinear patterns of official survey systems which laid on the land a geometric and dispersed settlement morphology of generally repetitious monotony.

There are, of course, important and symbolic exceptions to these generalisations. Rural New England, in particular, with its topographical diversity, colonial villages, irregular settlement pattern, English heritage and relative failure as an agricultural region, has acquired a degree of national status as a landscape ideal (Meinig 1979). Perhaps this is due also to the relative maturity of its cultural landscape. Certainly as one moves west into more recently settled areas, the countrysides of the east appear to gain in aesthetic appeal. Warkentin (1966), for example, has written of the westerner's view of southern Ontario as 'a serene countryside . . . with long views of sloping green meadows . . . closely spaced farmsteads holding substantial late nineteenth century brick houses . . . pulling all together into one integrated vista after another' (159). From the perspective of the flat, raw and sparsely populated prairie, Ontario, like many other areas of eastern North America must indeed give the appearance of the Garden of Eden. Yet, as Warkentin observes, Ontarians themselves have a poorly developed sense of their own countryside. While physical geography and settlement form may explain some of the weakness of aesthetic interest in the North American countryside, more fundamental are the values which have been associated with the settlement process itself. The prevailing images of rural North America are those of the pioneer, the family farm and the productive yet virtuous agrarian economy they worked to create. This produced not only a utilitarian and functional landscape, which was not readily identified with aesthetic ideals either by its creators or society at large, but also one which has become valued mainly in terms of the people who make their living from it.

The North American countryside ideal, therefore, has tended to value the settled rural landscape more as a symbol of agricultural progress and of bygone lifestyles than of aesthetic amenity. The farm and the small town rank high more as a desirable way of life than as a picturesque ideal. This is not to say that picturesque rural landscapes have not been attractive to North Americans. Certainly, as Chapter 3 will show, for almost two centuries urbanites have been attracted to the scenic amenities of the countryside. But the search for scenery has tended to look beyond farmscapes to more natural landscapes and to take comfort in the plenitude of open space to a degree that has long been impossible on the other side of the Atlantic.

2

THE ARMCHAIR
COUNTRYSIDE

That an idealised view of countryside should have emerged so rapidly from the turbulence of industrial revolution suggests that it was a natural and inevitable response to change. In a sense, of course, it was. The social and spatial disruption of urbanisation and the conditions of industrialism virtually guaranteed the emergence of sentimental and philosophical reverence for nature and rural life. Yet to become part of a civilisation's value system, ideals must be nurtured and diffused by the culture in which they are born. For a society increasingly separated from direct contact with land, nature and rural community, the main inspiration for the idealisation of the countryside has thus been the images and values presented by literature and art and, more recently, by an increasingly dominant range of mass media. It is these which have nourished and reinforced the mental images of the countryside, which have fabricated the mythology and romanticism within which nostalgic sentiment has flourished, but which have also whet the public appetite for real experiences of nature and country life.

The catalyst for this, as we saw in the previous chapter, was the rise of a literate and culture-conscious middle class – of a bourgeoisie attracted as much to the comfortable imagery of books and paintings as to direct experience of the countryside itself. From the mid-eighteenth century onwards, there was a rapid growth in the reading public, stimulated not only by the expansion of the educated classes, but also by improvements in the technology of publishing. The introduction of the steam-driven machine press in 1811, followed five years later by cylinder and rotary presses heralded the development of mass literary and artistic communication (Leavis 1932; Williams 1960). From this arose a rapidly expanding publishing industry, which set out to popularise the printed word, not only with cheaper and more accessible books, but also with a proliferation of periodical magazines and journals. To meet the literary demands of those who could not afford books, circulating libraries and book clubs began to appear. As Williams has observed, all this marked a radical change in the place of writers and artists in society. From being dependent upon the direct patronage of a tiny élite, they could now reach a wider public and thus become both more creatively independent and influential. The artist and the writer increasingly came to be seen, in Williams's words, 'as the guiding light, the revealer of truths' (Williams 1960: 33).

It was literature which dominated this cultural revolution. With the growth of mass education during the nineteenth century the written word acquired unprecedented power which it has retained to this day. Most of our armchair images and values associated with the countryside have come from the strong pastoral thread which runs through English literature. At one level this has served to explore some of the deeper questions about human existence which were discussed in the previous chapter. At another it has remained the literature of the intelligentsia. Yet the 'industrialisation of culture', as Horne (1986: 25) describes it, has blurred the distinction between the serious and the popular. The works of great writers have become popular classics, part of an expanding cultural world which also encompasses artistic and musical treatments of the pastoral theme. And while much of this remains within the realms of high culture, it is, to quote Leo Marx (1964), 'the region of culture where literature, general ideas and certain products of the collective imagination meet' (4). Pastoral imagery and nostalgia have spilled over easily into popular literature. And, with the arrival of new publishing technology and especially of electronic media – radio, film and television – in this century, the countryside ideal has been absorbed readily into mass culture. Indeed, the development of mass communication is just one facet of a consumer society in which images and values are marketed like commodities. Thus, as we shall see later in this chapter, the countryside has also become a cultural commodity, packaged by a publishing and communications industry and even used to sell goods and services.

The contrast between country and city is, to quote Marx (1964) again, 'an ancient literary device' (18). At times the contrast has been made in favour of the city. Certainly Medieval literature, read as it was by a small educated élite of nobility and clergy, paid little attention to rural life beyond the denigration of the peasantry and saw nothing inspirational about the natural world. And since then there has always existed a literary stream which has satirized country life and used rural folk for comic relief. The prevailing literary sentiment towards the countryside, however, has been a positive one (Keith 1974). At the heart of this is an enduring pastoral tradition.

This tradition has its roots in the classical pastoral; in the precise and artificial literary devices of the Theocritan *Idylls* and the Virgilian *Eclogues*. Theocritus adapted the popular songs and ballads of the Sicilian peasantry for the pleasure of sophisticated Greek society. His *Idylls* evoke an early sense of the countryside as an aesthetic ideal founded on 'the radiant splendour of a southern landscape' (White 1977: 25). In both poetry and art the landscapes of the Greek pastoral were romantic and idealistic. But the classical pastoral transcended actual landscapes to become an imaginative vision of a life of tranquillity and adundance, combining a magical view of nature with a bucolic fantasy of rural life. Central to this myth is the idea of a Golden Age both as present and future Arcadia. It is in Virgil's *Eclogues*, written amidst the splendour and turmoil of Augustan Rome that Arcadia becomes 'the magical Utopian vision' of an 'abundant land which needs no farming . . . a mingling of gods, shepherds and men in a departed golden world' (Williams 1973: 18). In this sense, the classical pastoral becomes mainly a poetic device – a convention for expressing a longing for a life of simplicity and pleasure.

The revival of the formal pastoral in Renaissance Italy and France provided the impetus for its adoption in the literature of Elizabethan England. Here in the poetry of Spenser, Sidney and Marlowe, in Shakespearian drama too, we find all the classical pastoral conventions. The poetic and dramatic expression of Arcadia, with its emphasis on romantic love in a pastoral setting was immensely popular with the Elizabethan court. The pastoral as a literary device continues into the next two centuries: in Milton's *Paradise Lost* for example and in the self-conscious revivalism of Pope and Dyer in the early eighteenth century. But from then on the classical form peters out. What does survive, by contrast, is a related but distinct stream of pastoral writing; a pastoral tradition which moves out of the world of Arcadian imagination and into the actual landscapes of the countryside (Marinelli 1971; Williams 1973).

GREEN LANGUAGE

Natural imagery is, of course, central to the classical pastoral. Some of the strongest images of the contrast between urban sophistication and rustic simplicity are furnished by Theocritan descriptions of the sights and sounds of nature. This is the oldest of pastoral themes: Arcadia as a natural paradise. Its reappearance in popular English literature in the Elizabethan pastoral, with its attention to natural beauty and the aesthetics of landscape, initiated what Williams has appropriately termed a 'Green Language' (Williams 1973); a literary and artistic celebration of nature and landscape which has become an extraordinarily popular element of Anglo-American culture over the last two or three hundred years (Finch and Elder 1990).

The poetry of nature

It is in the upsurge of English nature poetry during the late seventeenth and into the early eighteenth century through 'the conventional patter of poets who wouldn't be caught an hour away from town' (Thomas 1983: 251), that the popular idea of nature as the essence of the countryside began to emerge. The patter of poets, led by Pope, Dyer, Thomson, Cowper and Herrick, was, in fact, a virtual stampede to the patronage of country estates where poets who could lyricise the aesthetic pleasures of nature for a genteel and leisured readership were assured of a decent living (Humphreys 1964). The main attraction of this poetry was that it was replete with descriptions of nature and natural scenes, in contrast to the artificiality of the formal pastoral of the previous century. This shift was strongly influenced by the emerging ideas of the sublime and the picturesque, in which nature evoked both emotional and aesthetic response. James Thomson's poem, *The Seasons*, published in 1730 is widely accepted as one of the starting points in the emergence of a new poetic enthusiasm for nature. Although this long work covers, as Louis James (1989) has observed, 'the whole spectrum of eighteenth century attitudes to nature, from the theological to the scientific' (62), it is for its descriptive depiction of the natural tranquillity of the countryside that *The Seasons* acquired its popularity and influence. It became a bestseller on both sides of the Atlantic, as well as a favourite of heroines of popular novels (Hart 1950).

By the end of the eighteenth century, nature poetry had become one of the most popular sources of reading pleasure. To *The Seasons* we can add the retrospective lament for nature in Gray's *Elegy Written in a Country Churchyard* (which was as big a seller as *The Seasons*), Shenstone's portrayal of the natural harmony of the English countryside, Goldsmith's nostalgia in *The Deserted Village* for the nature of a golden rural age, and Cowper's gentle natural scenes (Keith 1974; Williams 1973). Poetry and art combined to foster a popular enthusiasm for scenic landscapes. Paintings of picturesque scenes became the dominant fashion for patrons of art, who rushed to buy the landscapes of Claude, Reynolds, Gainsborough, Wilson and Gilpin to decorate the walls of their country houses and urban villas.

With the emergence of pastoral romanticism the poetic celebration of nature came into its own. It is hard to exaggerate the influence of Wordsworth on the imagination of the early nineteenth-century reading public. It is not just that his poetry and his travel guides to his beloved Lake country were widely read, but that he and Dorothy also became icons of nature worship. Indeed, as Cavaliero (1977) has observed, 'Wordsworth has often been reduced in the popular mind to the status of a nature poet' (9). While Wordsworth stimulated enthusiasm for the wilder landscapes of the Lake District, it was the cosier version of nature set in the domestic landscapes of lowland England with which his generally genteel readership more often identified. Much of this ideal found its fullest expression in the 'Green Language' of the early nineteenth-century nature poets, dominated by Wordsworth himself and by John Clare (Williams 1973). Clare was the first in a long line of Victorian regional poets, describing in his sonnets the direct experience of nature in the landscapes of his beloved Fenland (James 1989).

By the mid-nineteenth century, however, a rapidly growing Victorian readership was indulging in a more sentimental version of the nature idyll. In Keats and Shelley, in Tennyson and Arnold, in the Rossettis and the Brownings, in Meredith and Hopkins, and in William Morris, mid-Victorian readers were presented with a vision of an English countryside in which woods, wildflowers, grassy-banks and birdsong were at the centre of the idyllic scene. As an escapist view of an English countryside in which nature symbolised a place where civilisation had not yet reached, this kind of poetry not surprisingly acquired a popular following amongst middle and upper classes struggling to relate to industrial progress. Tennyson's *English Idylls* epitomise Victorian enthusiasm for detailed yet evocative description of the natural sights and sounds of the countryside (Hunter 1984). But perhaps it is these immortal and much-memorised lines from Robert Browning which encapsulate the English idyll:

> O to be in England
> Now that April's there,
> And whoever wakes in England
> Sees, some morning, unaware,
> That the lowest boughs and the brushwood sheaf
> Round the elm-tree bole are in tiny leaf,
> While the chaffinch sings on the orchard bough
> In England – now!
>
> (R. Browning, 'Home-thoughts, from Abroad')

This poetic landscape painting was matched by the idyllic scenes of water-colourists like Millais and Holman Hunt, whose paintings of 'a still moment, usually in summer or autumn, heavy with sensations' (James 1989) were not only in great demand as prints to hang on living room walls but also illustrated volumes of poetry.

Although English nature poetry was set in the specific physical and cultural context of the English countryside, it was immensely popular across the Atlantic. Of course, the long arm of English culture had always reached deep into North American literary circles. But the great English nature poets were as well-known in Canada and the USA amongst the reading public at large as they were in Britain (Hart 1950). However, while nineteenth-century North American readers willingly soaked up the natural imagery of Wordsworth and Tennyson, they were also served by an impressive body of home-grown nature poetry. North American attitudes to nature, replete as they were with romantic and transcendental emotions about wilderness, guaranteed the development of the literary celebration of nature. Poetry has formed a significant part of this literature. Beginning with William Cullen Bryant in the 1820s and culminating in Robert Frost almost a century later, American nature poetry has helped to foster many of the prevailing images of the North American countryside. Bryant, 'the first great American poet' (Foerster 1950), with his prodigious output of poems with titles like 'A Forest Hymn', 'The Yellow Violet', 'Green River', 'To A Waterfowl', is widely recognised as the father of American nature poetry. By the 1830s he had attracted a wide readership and inspired the publication of 'a thousand amateur ditties and descriptive passages' by lesser poets, which celebrated the American wilderness (Clough 1964: 60). As in Britain these were matched by the prodigious output of painters of the American picturesque, led by Thomas Cole and the so-called Hudson River School of painting (Novak 1980). With other landscape artists like Thomas Moran and Frederic Church, Cole was the heart of the movement to bring the beauties of the American landscape into the homes of urban society.

Bryant was in the vanguard of a growing stream of nature poetry which, by the 1860s, was in full flood. It was dominated by the 'huge output' of the New England poets: Whittier, Longfellow, Holmes, Lowell, whose 'easily read verses ... invited a large and appreciative audience that read not to discover new ideas but to have old ones repeated and confirmed' (Stauffer 1974: 160). To their largely simple descriptions of nature in the familiar settings of the New England landscape, was added the more inspirational nature poetry of Emerson, Dickinson and Whitman. Emerson expressed his transcendental ideals as much in poetry as in prose. His poetic celebrations of the restorative powers of nature, where 'the gods talk in the breath of the woods', were widely published and did much to make him the icon of mid-nineteenth-century nature worship. Neither Emily Dickinson nor Walt Whitman were exclusively nature poets, but it was for their reflective and celebratory poems about nature and civilisation that they acquired much of their reputation (Kazin 1988).

Poetry clearly played an important role in engendering and reinforcing popular images of nature in the nineteenth century. Although it was not the most widely read form of literature, the nineteenth-century reading public on both sides of the Atlantic was exposed to nature poetry firstly in numerous published volumes (which, in the case of the great names, often went to multiple editions), and, secondly and more generally, through the widespread publication of poetry in popular periodicals and magazines. Moreover, if there is any poetry which has been retained in today's popular culture, more often than not it is the 'classics' of the nature tradition of the last century – lines from Wordsworth, Tennyson, Bryant and Longfellow lingering from schoolday memorisation. This is not to say that nature poetry is an entirely nineteenth-century phenomenon. It is a tradition kept alive in American poetry by Robert Frost, Robinson Jeffers and Wallace Stevens. Frost is, without doubt, the best known of all American poets. His enduring popularity, which extends well beyond the borders of the USA, owes much to his portrayal of the rustic, wooded landscapes of New England, of scenes replete with birches, apple trees and snow-filled woods.

Plate 2.1 The American picturesque: *View on the Catskill, Early Autumn* by Thomas Cole, 1837

Much the same can be said of Frost's English contemporaries – Edward Thomas and the lesser poets of the so-called Georgian movement of the Edwardian years (Keith 1974), who saw feeling for nature as an integral element of their nostalgia for the old rural England.

Modern poetry has moved a long way from its pastoral roots. Yet nature has remained a remarkably durable poetic theme. In part this may be because for much of society nature and poetry, in the often enforced memorisation of the likes of Frost and Wordsworth, have been synonymous since childhood. But it is also because nature remains such an accessible theme, especially in an age of increasing environmental sensitivity.

Literary naturalists

While nature poetry has helped to foster an idyllic image of the countryside, it is only part of a much broader tradition of nature writing. At the heart of this is a mix of non-fiction literature ranging from the observations of amateur naturalists to the eco-centric essays of modern environmentalists. Its roots lie in the emergence of natural history as a branch of literature during the early eighteenth century (Johnson 1966). Such was its popularity by the middle of the century that one contemporary observer was prompted to note that works on natural history sold more than any other kind of book in England (Thomas 1983). Beginning with John Ray's classification at the end of the seventeenth century and culminating in Gilbert White's famous letters from Selborne almost one hundred years later, the writings of countless amateur naturalists promoted in eighteenth-century society the idea of nature study as 'intellectually rewarding, spiritually edifying and aesthetically gratifying' (Thomas 1983: 283). Yet these were not scientific discourses, rather 'gracious and humane' (Johnson 1966: viii) descriptions of nature which set out to stimulate the interest of the ordinary reader.

It is Gilbert White who is generally regarded as the pivotal figure in this kind of literature. His *Natural History and Antiquities of Selborne*, was first published in 1789 and has since passed through almost one hundred editions (Johnson 1966). In an age which saw nature through picturesque and romantic filters, the parson–naturalist's descriptions of the 'natural curiosities' of his cosy corner of southern England were an instant hit. White was the first in a long line of literary naturalists who had a powerful influence on attitudes to nature and the countryside on both sides of the Atlantic. In Britain they tended to be drawn from the ranks of the country clergy and gentry, whose education and lifestyle was particularly suited to such pursuits. Among the most notable of these was John Knapp, the son of a rector, who transcribed his daily walks around his country estate into *The Journal of a Naturalist*, which was first published in 1829 and went to three more editions during the next ten years (Johnson

1966). Others included the eccentric landowner Charles Waterton and Charles St. John whose combination of descriptions of field sports and nature in the Scottish Highlands mirrored the contradictions in Victorian upper-class attitudes to nature. Waterton and St. John were part of a flurry of this kind of publication between the 1840s and 1880s. The Reverend J. G. Wood's *Common Objects of the Country*, published in 1858, sold 100,000 copies in a week (Thomas 1983). Equally popular was Philip Gosse's *A Naturalist's Rambles on the Devonshire Coast* (1853), and the *Prose Idylls* of yet another parson–writer Charles Kingsley (Johnson 1966).

While English amateur naturalists were entertaining their readers with descriptions of the flora and fauna of a familiar and domesticated country-side, their American counterparts were exploring whole new natural regions. Although they were strongly influenced by Gilbert White and the general eighteenth-century tradition of English nature writing, post-independence Americans, as we saw in the previous chapter, quickly developed an enthusiasm for the exploration and description of the natural features of their own landscapes. Moreover, early accounts of the natural wonders of the New World, such as the letters of De Crevecouer and Audubon's famous ornithology (1827–38), drew as appreciative a read-ership back across the Atlantic. But the real growth in American nature worship came with the rush of amateur naturalists, writers and artists to explore the wilderness and to transcribe their observations into print and plate. During the 1830s and 1840s middle-class urbanites could enjoy the natural world from the comfort of their homes and libraries through the pages of Washington Irving's, *A Tour on the Prairies* and William Cullen Bryant's descriptions in the *Evening Post* of his trips to the Palisades, the Delaware Water Gap and other natural features. Especially popular were the writings of Charles Lanman, whose lively travel essays, first published in magazines, were collected into 'summer books', which were very much in fashion in the 1840s and 1850s (Huth 1957: 82). By the 1860s and 1870s 'those who communed with nature at their writing desks flooded newpapers and magazines with Arcadian essays' (Schmitt 1969: 16), while explorers like John Wesley Powell made the remote wonders of the west accessible to easterners through articles in *Scribner's Magazine* (Finch and Elder 1990). Scores of similar articles appeared in *The Atlantic* and *Harper's Weekly*. While much of this literature described natural scenery, there was also a steady stream of publications on the standard subjects of amateur naturalists: wildflowers, trees, insects, birds. Eighteen of the books by the great nature guru John Burroughs had bird titles. These sold over a million copies between 1880 and 1933 (Schmitt 1969).

The fashionable interest in nature was also propagated by a growing wilderness and outdoors recreation movement which, as Chapter 5 describes in detail, reached its peak in the early decades of this century. (Nash 1967; Schmitt 1969). The Reverend William Murray's *Adventures in the Wilderness or Camp-Life in the Adirondacks* (1869) was among the first in a long line of books extolling the virtues of the direct enjoyment of nature. In the writings of the Canadian outdoors protagonists Ernest Thomas Seton and Sir Charles D. Roberts, descriptions of nature excursions took on a decidedly anthropomorphic and sentimental tone which drew much criticism from other naturalists but sold lots of books (Schmitt 1969: 46). This was part of a turn-of-the-century nature revival movement which was also popularised by writers such as Dallas Sharp and Gustav Stickley. One of what Schmitt has called 'literary commuters', Sharp wrote numerous essays in *The Atlantic Magazine* promoting the ideal of suburban living where the commuter could stay 'contentedly within the sound of the dinner horn, glad of the companionship of his bluebirds, chipmunks and pinetrees' (22). A leading figure of the arts and crafts fraternity, Stickley wrote mainly in *The Craftsman* magazine of the virtues of country living in the 'free spaces of nature' (in Shi 1985).

In the writings of the proponents of outdoor recreation and country living we can recognise an important shift in the literary treatment of nature, which has had significant implications for the armchair appreciation of the countryside. By moving away from the detached objectification of many of the earlier literary naturalists to a more personal engagement with nature, writers like Seton and Stickley established in the minds of their readers sentiments about the natural world which could extend beyond its vicarious appreciation in the printed word to its direct physical enjoyment. This was a shift which occurred on both sides of the Atlantic. In Britain it emerges initially and most powerfully in the writings of Richard Jefferies. In contrast to the parson–naturalists and gentry–naturalists who preceded him, Jefferies was a struggling journalist, the son of a failed Wiltshire small-holder, a displaced countryman writing for a predominantly urban readership (Keith 1974). From his home in the London suburb of Surbiton he produced numerous newspaper articles, essays and books in which he revived the detailed memories of the countryside of his youth (Williams 1966). Although, as we shall see later in this chapter, much of his writing was concerned with country life, he wrote a good deal about nature. In *The Gamekeeper at Home; or, Sketches of Natural History and Rural Life* (1878), *Wild Life in a Southern County* (1879) and *Nature Near London* (1883) he exposed his readers to the detailed pleasures of nature in an accessible landscape. That these and indeed most of his books initially appeared in serialisations or essays in popular magazines such as the *Pall Mall Gazette* and the *St. James's Gazette*, and even in daily newspapers, has much to do with his instant and sustained popularity (see Mabey 1983).

During the Edwardian and inter-war years the kind of intimate celebration of the nature of the English countryside which characterised Jefferies' writing was a central element in the recurrent rural nostalgia of the period. Key figures in this were W. H. Hudson and Edward Thomas. Hudson's descriptions of his rambles around the countryside published in *Nature in Downland* (1909) and *A Shepherd's Life* (1910) combined the keen eye of a serious amateur naturalist with the sentiment of a nature lover. Perhaps, as Keith (1974) has observed, his popularity was more due to his style than his subject matter. But then the very intent of this literary genre was to inspire enthusiasm for nature. This is particularly evident in Edward Thomas, who wrote volumes of essays about nature in southern England long before he turned to poetry. In *The Woodland Life* (1897), *The Heart of England* (1906), *In Pursuit of Spring* (1914) and at least half-a-dozen books along similar lines, he produced rapt and detailed descriptions of nature and country life for, by his own admission, 'the villa residents and the more numerous who would be villa residents living in or on London' (in Keith 1974: 194).

That this was a literature written principally for an urban market only served to heighten its sense of escapism, but of an escapism in which nature was portrayed in familiar and accessible contexts. Nature, too, especially when set in the cosy landscapes of lowland England, served to define much of the essence of the countryside ideal, and indeed of the national character as a whole. This is readily apparent in the volumes of country writing which appeared in the inter-war years; in, for example the almost idyllic world of Henry Williamson and in the publications of a new breed of literary naturalists like H. J. Massingham, H. E. Bates and Alison Uttley, and farmer-naturalists such as Adrian Bell and A. G. Street, who combined nature description with a general nostalgia for country ways. The popularity of these writers as well as this type of literature has continued to the present day in both the continued re-issuing of the works of earlier nature writers and in contemporary exponents such as Richard Mabey and Gordon Benningfield. It is also a view of nature which has been communicated through the long tradition, stretching back to the last century, of 'country diaries' in the more serious newspapers and in the nature columns of *Country Life* and other rural magazines.

Nature and environment

Much of the contemporary popularity of nature writing is sustained by the growing public concern about environmental degradation. The valuation of nature as the refuge for the human soul and as the defence against industrialism, as we saw in the previous chapter, stretches back to the romantic period. Wordsworthian ideals must have pervaded educated minds even if they were generally attracted by a simpler idyll. And, of course, with the rise of transcendentalism, nature acquired a significance with which a society grappling with the ambiguities of material progress readily identified. Emerson was a towering social, as well as literary figure in mid-nineteenth-century America, lecturing, writing essays in fashionable magazines and, most notably with *Nature* and *The Journals*, publishing books which were virtually mandatory in the libraries, public and private, of respectable society on both sides of the Atlantic. His contemporary, Thoreau, although not particularly widely read in his own time, stands at the centre of the back-to-nature and environmental movement. By the 1880s *Walden* had become a best-seller, disseminating in its accounts of life in the Concord woods, a sense of materialist threat to both nature and humanity which bridged the gap between nature description and philosophy in a way which made sense to the average reader. It was this blend which drew late nineteenth- and early twentieth-century readers to John Burroughs and John Muir, arguably the two most well-known figures of American nature writing (Finch and Elder 1990). Both wrote voluminously for a large and appreciative magazine and book clientele – Burroughs adding to his naturalist observations a growing sense of the benefits of nature for civilisation; Muir translating his transcendental ideology into a populist campaign for wilderness preservation and enjoyment – and in the process did much to kindle interest in the natural world even amongst those who rarely ventured farther than an urban park. Burroughs's famous article, 'What Life Means to Me', published in a 1906 issue of *Cosmopolitan* magazine, illustrates the kind of writing which drew public attention:

> Oh, to share the great, sunny, joyous life of earth! to be as happy as the birds are! as contented as the cattle on the hills! as the leaves of the trees that dance and rustle in the wind! as the waters that murmur and sparkle the sea!
>
> (in Shi 1985: 186)

We can discern a similar, albeit less explicitly articulated thread in the British nature writing of the same period. Much of Richard Jefferies' immediate popularity and influence can be credited to the underlying sense of nature's value to modern humanity which pervades a good deal of his writing (Mabey 1983: 22–3). The same kind of feeling for nature and the countryside as a disappearing world, runs through the work of W. H. Hudson and Edward Thomas. Indeed Hudson was active in the emerging nature conservation movement of the times (of which more in Chapter 6). As this movement grew on both sides of the Atlantic so too did the public enthusiasm for reading about nature. In inter-war Britain, the popularity of nature themes was fostered by the huge output of writers such as H. J. Massingham, who published some thirty books on nature and countryside topics between 1921 and 1952. 'My theme,' he wrote in his introduction to *Through the Wilderness* (1935) 'is the relationship between man (*sic.*) and nature in our country, its fruitfulness and the disastrous consequences of disturbing it' (in Keith 1974: 238). Massingham's combination of nature description and conservational philosophy, coming as it did from a reactionary who totally rejected the ideology of industrial progress, touched a responsive chord in British society.

Across the Atlantic, North American readers were absorbing similar (although contextually quite different) treatments of nature in the writings of leading inter-war conservationists (Nash 1967). Yet, the real surge in public appreciation of nature as an essential element of environmental awareness came with the new naturalism of the seventies. The writings of Thoreau, Burroughs and Muir experienced a major revival. Aldo Leopold's *Sand County Almanac*, which was largely ignored when it first appeared in 1949, quickly became, with its blend of nature description and environmental ethics, the bible of the new environmentalism, selling over 270,000 copies in 1973 alone (Shi 1985). Leopold's philosophy spawned a new generation of nature publications. Indeed on both sides of the Atlantic, books and articles, indeed whole magazines on nature and wildlife have now become an important source of income for the publishing industry. Amateur naturalists and back-to-the-landers find a ready market for books which portray the virtues of 'ecological simplicity' (Shi 1985: 263). Much of the contemporary interest in the countryside is stimulated by the huge output of publications on nature conservation. This is especially true of Britain where naturalists, professional and amateur, like David Bellamy and Richard Mabey, often supported by agencies and organisations such as the Nature Conservancy, the Countryside Commission and Common Ground, purvey an implicitly educational message to an apparently receptive and appreciative public. Television has come to play a central role in this on both sides of the Atlantic, with a steady diet of nature programming which has become a staple of small screen entertainment.

COUNTRY LIFE

Any list of the most popular television programmes of the past twenty years would have to include *All Creatures Great and Small*. Running into several series and numerous repeats, and watched by millions around the English-speaking world, this adaptation of James Herriot's anecdotes of his experience as a veterinary surgeon in the Yorkshire Dales has drawn a huge following of viewers to its images of country life. Despite its autobiographical authenticity, its enduring attraction is its nostalgic veneer. Set in the period between the 1930s and the 1950s, it is close enough to the present to be selectively recognisable by its largely middle-aged and elderly audience, but historical enough to perpetuate some of the mythology of village life and country ways. *All Creatures Great and Small* has all the elements of rural nostalgia, from rustic farm characters and earthy agricultural customs to the eccentric squirearchy and the parochial village community. It stands at the end of a long tradition casting back to the rural 'Golden Ages' of the immediate past (Williams 1973), in a social pastoral which looks not so much to nature itself as the source of rural bliss but to the old rural ways and communities in harmony with nature and the land as the nostalgic antithesis of urban life.

The old order

The underlying attraction of Herriot's stories is that they portray the vestiges of an old social order. It is not just that his countryside contains the quaint remnants of pre-industrial ways of living, but that it stands for a world in which social relations were defined by a benign rural class system. This has been a remarkably durable view of English country life. It has been nurtured and perpetuated by a literary tradition centred in the first instance on the country estate but ultimately extending to the hierarchical society of provincial England.

The country estate and the leisured lifestyles which it supports has long been a popular literary subject, and one which has done much to fashion sentimentalised images of country life. By the beginning of the eighteenth century, the literary celebration of the pleasures of life on a country estate which had first appeared in Elizabethan pastoral poetry, had become a prominent and highly popular literary theme (Humphreys 1964). The Horatian ideal of the 'happy man' in the peace and plenty of his country estate, revived by Pope, Cotton and Dryden and countless lesser poets in the pay of the gentry, was further popularised by a large volume of literature which focused on the sporting life of the country gentleman. Gay's poem, *Rural Sports*, Somerville's *The Chase* and *Field Sports*, and the hunting characters in novels like Fielding's *Tom Jones* helped to foster the perception of country life as a carefree idyll. Even the satirical overtones of Fielding and, of course Sheridan, did little to diminish the life of the country gentry as the filter through which country life as a whole was viewed.

Given that it was the gentry, and the growing number of social climbers who aspired to that status, who made up the bulk of eighteenth-century readership, the popularity of country house literature is hardly surprising. By the beginning of the nineteenth century, however, the emphasis had shifted from the portrayal of the simple pleasures of country living, to the social intricacies of a squirearchy operating within the comfortable boundaries of the country house and the local country town. Much of this had to do with the rise of the novel as the popular literary form. Thus it was the gentrified provincial world of Jane Austen and then, by the early Victorian years, of Mary Howitt, George Eliot, Elizabeth Gaskell, the Brontes and Anthony Trollope, which became such a popular subject. Austen's novels are about the landed society, the 'high bourgeois society' as Raymond Williams (1973: 14) has called it, which at the end of the eighteenth century was busily transforming the English countryside into a landscape of country estates within which the essential morality of Victorian England would be determined (Short 1991). Fifty years later, Trollope, in his Barchester novels, portrayed the maturation of this society, with its clerical aristocracy and lesser gentry, 'its constant county hunt and its social graces' (Williams 1972), into the dominant force in English country life. This, of course, is precisely what his respectable readership, which consisted mainly of ladies of leisure and social position, wanted to read about. It satisfied their image of old, provincial, upper-class England; of a philanthropic landed gentry, surrounded by the architectural trappings of success and ruling over a tranquil countryside.

That the attractiveness of this image, and especially the fascination for country house society, has persisted into our time owes much to the enduring popularity of the Victorian novelists and to their status as literary classics. Yet it is influenced also by the continued centrality of the country house as a setting for more recent literature. From the turn of the century right through to the outbreak of the Second World War, English literature and drama, both popular and serious, has retained the country house as a setting 'for a more general social drama' (Williams 1973: 249). In the interwar years in particular, in novels like those of P. G. Wodehouse and Evelyn Waugh, and in the plays of Noel Coward, the country house set, with their house-party weekends and eccentric lifestyles were a popular source of entertainment. The country house also became the preferred setting for the detective stories of Dorothy L. Sayers and Agatha Christie, as well as for numerous romantic historical novels. All these, of course, remain enormously popular amongst readers, theatre-goers and television viewers today.

Much of the increasingly nostalgic attachment to images of country house society that have been fostered by literature over the past two or three centuries stems from the perception of the gentry as the central symbol of the old rural social order. In this sense, country house literature is part of a more general culture of sentimentalising traditional English rural society; of idealising country folk and village communities as contented and harmonious beneficiaries of the rural hierarchy. In its early form, at the turn of the eighteenth to the nineteenth century, it is a lament for the destruction of a contented peasantry by agricultural improvement in Goldsmith's famous poem, *The Deserted Village* (1770), and then in William Cobbett's *Rural Rides* (1830). Despite his radical reputation, Cobbett's concern for the condition of rural labourers at the end of the eighteenth century, 'led him back to the early part of the century in a nostalgic vision of a time when "each had his little home," when they lived in the same cottage and worked for the same master all their lives' (Keith 1974: 71). Yet while Cobbett was looking back to another 'Golden Age', more conservative writers like William Cowper, Jane Austen and Robert Southey were fabricating the harmony of contemporary rural society.

By the early Victorian years Mary Mitford and George Eliot were providing their readers with reassuring images of traditional village England, full of contented country characters and 'sheltered cosiness' (Keith 1974: 87). This, of course, was also the England of Constable, who in paintings like *The Cornfield* (Plate 2.2) had already transformed the picturesque tradition into a portrait of an English rural garden populated by a contented peasantry engaged in rustic tasks. The sentimentalised portrayal of country folk and village life became a central element of

Plate 2.2 The old order of rural England: *The Cornfield* by John Constable, 1826

the mid-Victorian idyll (Treble 1989). The epitome of this image were the paintings of Myles Birket Foster and his disciple, Helen Allingham, whose water-colours of happy country folk set against a picturesque background of villages and cottages initiated a whole tradition of sentimental artistic portrayals of country life which have persisted to this day (Plate 2.3).

As the nineteenth century progressed the literary celebration of the old rural order became increasingly nostalgic and romantic. Essays and books lamenting the passing of 'good old' rural England were commonplace by the 1890s, and by the turn of the century rural nostalgia was an important resource of the publishing trade (Cavaliero 1977). One stream of this consisted of the intellectual anti-modernism of Ruskin and Morris. While this certainly contributed to the idealisation of traditional country life, it was quite different from the sentimental outpourings of the conservative writers who by the Edwardian years were trumpeting the virtues of the true England: a Constable landscape of cottages and villages, farms and fields, inhabited by a contented society which still clung to the old ways and the old rules. This was a reassuring myth for upper and middle classes confronted with the rise of socialist and modernist values. They read with enthusiasm the conservative nostalgia of the Poet Laureate Alfred Austin, who in 1901 published an account of a trip through England entitled *Haunts of Ancient Peace*, they absorbed Kipling's gentlemanly nostalgia for rural and historical England in *Puck of Pook's Hill* and above all they savoured the sense of the timelessness of a countryside viewed through the eyes of English gentlefolk. National character, as we saw in the previous chapter, became associated with ancient rural virtue, a sentiment which was popularised not only by poets and novelists, but also by conservative historians (Wiener 1981). The romantic notion that the old order of the countryside was England's true heart was repeated in several books with similar titles, notably *The Heart of the Country* (Ford Madox Ford 1906), and two entitled *The Heart of England*, by Edward Thomas (1906) and Hilaire Belloc (1913).

This idea grew in popularity during the inter-war years with a proliferation of books, articles and radio programmes on English life and heritage which largely meant anything to do with the countryside. One of its most popular exponents was H. V. Morton, a rural myth-maker of the first order, who in articles in the *Daily Express* and then in a book, *In Search of England* (which ran to 12 editions by 1936) set off to find the true English way of life in the countryside. The romanticised view of country life was kept alive, too, by a steady stream of fictional literature. Nature, landscape and country folk, usually in a regional setting, combined to create an element of social fantasy in the Sussex novels of Sheila Kaye-Smith, in Hugh Walpole's *The Herries Chronicle*, Mary Webb's series of Shropshire novels, T. F. Powys' stories set in Dorset villages and in Francis Brett Young's manor house settings (Cavaliero 1977; Keith 1988). Although

Plate 2.3 The mid-Victorian rural idyll: *Cottage at Shere* by Helen Allingham, 1856

they wrote from direct experience of the countryside, theirs was largely a vision, as Keith has observed of Mary Webb's books, of a 'timeless realm, pre-First World War and pre-motor car' (Keith 1988: 130). The fabrication of this country life myth reached such proportions that Stella Gibbons was prompted to write *Cold Comfort Farm*, a stark and depressing, yet widely read anti-pastoral aimed satirically at the rural sentimentalisation of the times (Williams 1973).

The focus of much of this sentimentalisation of English country life has been the village. Hilaire Belloc believed that the true heart of England was to be found in the village. Village life and landscape were as symbolic of the old social order as the country estates of which they had so long been a part. Victorian readers, as we have seen, had already been attracted to this view. In this century the village has acquired almost mythological status as the archetypal English community. To some extent this stems from the way the village has been treated in country house literature, that is, as a peaceful yet dependent extension of the structure of landed society. The village is also an essential part of the timeless rural world of the regional novel. Yet much of our nostalgic attachment to the village has been fostered by a literature which directly celebrates village life. There is no more popular exponent of this than the pseudonomous 'Miss Read', whose many paperback editions have been best-sellers for the past twenty years. The attraction of her village romances is that they are set in a vague time-frame somewhere between the 1930s and the 1970s, in an immediate past with which her readers can identify, but not so immediate that the realities of modern village life need mentioning. Miss Read's village retains all the gentility of the old social order and a sense of quaintness which can be also found in the anecdotal treatment of village life popularised by such writers as H. E. Bates, Laurie Lee and James Herriot. Bates's Larkin family series (recently serialised on television) and Laurie Lee's, *Cider With Rosie*, with their caricatures of country folk and social relations, have done much to perpetuate the myth of the village and the countryside in general as the last remnant of a happier way of life in which all classes co-exist in tolerant harmony. Perhaps the best example of this is the long-running radio series, *The Archers*, 'an everyday story of country folk', which has been the filter through which generations of urban listeners have perceived village and country life.

'One of the causes of the charm of an English village,' wrote Ditchfield (1898) 'arises from their sense of stability. Nothing changes in country life' (2). For Ditchfield, however, the charm of villages lay in their appearance rather than their social life. His descriptions of village architecture mark the beginning of an expanding area of publishing which reflected a growing turn-of-the-century interest in the exploration, both actual and vicarious, of the English countryside. By the 1930s the publishing company Batsford began producing a series of illustrated volumes on the villages and hamlets of England. These have been re-issued in recent years but have also been joined by a new generation of 'village books', which, as we shall see later in this chapter are part of a growing industry of countryside publications. What both the Batsford series and its contemporary counterparts have in common is their portrayal of the vernacular architectural charm of England's most picturesque villages. Theirs is not a representative survey, but rather a selective depiction of the village as a traditional place. Perhaps the best illustration of this is a recent photographic essay entitled *Old English Villages* in which there is a complete absence of people or vehicles. 'My aim,' writes the photographer–author, 'while taking the photographs was to create a timeless atmosphere, so that the villages could be seen as they may have been at the turn of the century, before the petro-chemical age changed most of England beyond recognition' (Perry *et al.* 1986: 156).

In many respects the realities of country life that, as Williams (1973) has put it, 'were scribbled over and almost hidden from sight by what is really a half-educated and suburban scrawl' (258) have become steadily more appreciated and understood by educated society. Certainly the conservative filter through which country life was viewed by generations of readers up to the 1940s would now be regarded by many as at best quaint and at worst downright ignorant. Yet that literary perspective is not easily dislodged from English culture. There remains in the popularity of its contemporary exponents and, even more obviously in the re-publication and television dramatisation of the earlier classics of the genre, a fascination for the old rural order which is perhaps more than just historical curiosity. England remains a profoundly conservative country, uniquely respectful of its nobility and its gentry. Images of the old order fuel this conservatism and, by the same token, help to foster both the middle-class attraction for country living and much of the preservational sentiment towards the countryside.

Farm and frontier

Americans, or at least white middle-class Americans, have thoroughly embraced the conservative image of rural England. The classics of English country house literature, as well as the more popular romanticisations of the old rural order, have always had a large American readership, while their film and television adaptations have also acquired an enthusiastic audience. The absence of a direct counterpart either in the literature or the social history of their own countryside has much to do with this. To be sure, there are parallels to be found in the historical romances of the Southern plantation. For the most part, however, American literature in particular has romanticised a simpler and more egalitarian social order which has centred on the twin pillars of the American democratic ideal: the farm and frontier.

Although the idealisation of agrarian society is associated most directly with Jefferson, much of the persistence of the agrarian myth in the American mind can be attributed to fictional literature and, more recently, to the portrayal of farm and country life in film and television. Indeed, as Shi (1985) has observed, as commercialism quickly overtook the Jeffersonian agrarian republic, the ideals of simple, agrarian virtue 'retreated from the realm of public policy into the realm of rhetoric' (101). By the second quarter of the nineteenth century agrarianism had become for most urban Americans a romanticised image of farms and farming communities populated by simple, decent country folk living in harmony with nature, the land and each other. It is an image which runs through ante-bellum poetry and art; Bryant, Whittier, Longfellow and Cole matching their nature worship with eloquent impressions of agrarian rusticity. It inspired genteel novelists like Sarah Hale, whose moralising tale in *Northwood* (1827) of a young man returning from an experiment with city living to the simple pleasures of farm and small-town life became a best-seller not only in its own time but for twenty years afterwards (Hart 1950; Shi 1985). With Catherine Beecher, Hale was a leading figure in the cult of domesticity. Her association of the domestic ideal with the virtuous simplicity of farm and small-town life drew a genteel and largely female readership, especially through her editorials and articles in the pages of the *Ladies' Magazine* (Shi 1985).

From these ante-bellum beginnings, the literary mythology of agrarian simplicity came to dominate American attitudes to the countryside throughout the nineteenth and indeed well into the twentieth century. While, as we saw in the previous chapter, at a philosophical level it was fostered by the anti-progressive ideology of Thoreau and Norton, at a more popular level it was disseminated by a wide range of fictional literature. The nature poetry of Whittier, Whitman and Frost is suffused with images of farm and country life. Magazines from the *Lady's Book* to *Harper's* maintained a steady stream of stories set in the cosy world of the farm and the small town. By the early twentieth century novels about country folk, which, 'satisfied many Americans yearning for the rural ways they had left in searching out a living in the big cities', dominated the best-seller lists (Hart 1950: 205). Among these were Edward N. Westcott's 1898 novel of homespun farm life, *David Harum*, which by 1904 had sold 750,000 copies; John Fox's *The Little Shepherd of Kingdom Come* (1903); Kate Douglas Wiggin's *Rebecca of Sunnybrook Farm*, also published in 1903, which became one the most popular American books of all time; and finally Gene Stratton Porter's country life idylls, *Freckles*, *A Girl of the Limberlost*, *Harvester* and *Laddie*, each of which by 1946 had sold over a million copies (Hart 1950). That the last two of these writers attracted a readership of both adults and children only serves to highlight the element of escapism which was such an important part of the popular mythology of rural life.

As Roy Meyer has observed, much of the twentieth-century farm literature has focused on the Middle West (Meyer 1965). The agricultural heartland of modern America, but also once the agricultural frontier, the Middle West has been fertile ground for the literary portrayal of the combined virtues of pioneer spirit and farm life. At least one hundred and forty novels dealing with Middle Western farm life were published between 1891 and 1962. During the inter-war years, in particular, novels celebrating the conservatism, individualism and anti-urbanism of farm society, such as Phil Stong's *State Fair*, and Edgar Watson Howe's *The Story of a Country Town*, were among the nation's most popular fiction. Some novelists of this genre, like Willa Cather, Margaret Wilson, Edna Ferber, Louis Bromfield and Josephine Johnson became Pulitzer prize winners (Meyer 1965). After the depression, and particularly with the broadening of American literature in the post-war years, the farm idyll waned somewhat as a novelistic theme. Yet sympathy for the simple farm family is invoked by Steinbeck's *Grapes of Wrath*, while the same author's later book *East of Eden* is permeated with Jeffersonian agrarian sentiment (Maguire 1991). Moreover, as television has increasingly taken over the role of shaping popular values, the romanticised portrayal of farm and country life has attracted a large audience. The best example of this must be the long-running television series *The Waltons*, which dominated the viewer ratings during the 1970s with its sentimental portrayal of a poor rural family surviving the Depression with the moral fortitude, innate wisdom and sheer old-fashioned decency of country folk.

Whether or not the frontier really shaped the character of American society as Frederick Jackson Turner (1920) argued, its treatment in literature, film and television has certainly established it as one of the prevailing American myths. The heroic literary figure of the pioneer and the backwoodsman, of course, goes back to James Fenimore Cooper's *Leatherstocking Tales* and to the dime novels which glorified the exploits of frontier heroes such as Daniel Boone and Kit Carson (Maguire 1991). But it is the image of the taming of the west – of farmers and ranchers carving the expanding nation out of the wilderness – which has captured not only the American but also much of the world's imagination. As Karolides (1967) has observed, since the closing of the frontier in 1900, the early American drama has been re-created again and again in a variety of forms, but especially in prairie trail adventures like Emerson Hough's *The Covered Wagon* (1922), and A. B. Guthrie's, *The Big Trail* (1947) and *The Way West* (1949), and in stories of prairie farmers like Willa Cather's *O Pioneers!* (1913), C. R. Cooper's *Oklahoma* (1926) and Phil Stong's 1937 novel *Buckskin Britches*. These, however, are really part of the broad tradition of the Western, at the centre of which is the cowboy. Starting with the classics by Owen Wister (*The Virginian*, 1902) and Zane Grey, the greatest of all writers of Westerns, who by 1922 had published 16 novels, of which twelve had already been made into movies, the image of the cowboy as the 'last romantic figure upon our soil' became the stuff of countless novels (Maguire 1991: 439). Yet the most powerful images of the old west as one of the defining symbols of American country life have come to us through the Westerns of film and television. The Western, observes Short, 'embodies the general myths of wilderness and countryside' (Short 1991: 178). The classic Western film is really about sex, shoot-outs and the slaughter of Indians, but all this functions against the backdrop of the settling of the west and the establishment of raw and simple rural communities. This latter image has come through most explicitly and influentially in serialised television Westerns like *Bonanza* and *Little House on the Prairie*. In contrast to most of the great Western films, these small-screen versions have presented a romantic version of frontier life, which has clearly been popular with generations of viewers (Gitlin 1983).

The view from within

To argue that the armchair image of country life has been shaped only by the romantic perspectives of urban writers would be to ignore an important body of literature which has been written by those who either live in the country or who know it intimately. From these has come a more serious and accurate account of country matters, in which nostalgia blends with understanding to become a reflective appreciation of rural folk and rural ways. William Cobbett is generally regarded as the father of this perspective in rural literature. His nostalgia for a pre-enclosure rural order notwithstanding, Cobbett's detailed observations drawn from two decades of touring the countryside on horse-back set out to reveal the contemporary social and economic conditions of rural England. Although *Rural Rides*, first published in 1830, is now considered one of the classics of country writing, its impact on early Victorian readers is harder to determine. Certainly there were others at the time, like John Clare, who were concerned with presenting a true picture of, and in the process stimulating an educated interest in, ordinary country life (see Keith 1973). Yet Cobbett's was at once a conservative and a radical treatise, popular for its opposition to progress yet suspect for its exposure of rural poverty. It was thus his detailed descriptions of everyday rural life – of a surviving folk culture – which attracted contemporary readers.

This particular way of looking at country life, however, is much more a late than an early nineteenth-century development. By the 1860s William Barnes was attracting a considerable following with public readings of his dialect poetry portraying the traditional rural society of his native Dorset (Keith 1974). But it is the two giants of Victorian country writing, Richard

Jefferies and Thomas Hardy, who are most clearly identified with the late nineteenth-century interest in country matters. Jefferies was much more than the nature writer described earlier in this chapter. In numerous newspaper articles, essays and books written between 1872 and 1887, he revived the detailed memories of his youth and of his summer visits to his native Wiltshire. His was a realistic portrait of rural society, tinged with political judgements on the social hierarchy, yet sufficiently celebratory of traditional country folk and ways to kindle a growing urban interest in the intracies of country life. In *The Gamekeeper at Home, Round About a Great Estate* and *Hodge and His Masters,* each of which was first serialised in the *Pall Mall Gazette* or the *Evening Standard,* Jefferies wrote knowledgeably and affectionately about a wide range of country topics. Wiener (1981) has observed that his writings marked 'a watershed between an older rationalisation of the rural status quo, and an emerging urban-centred ruralism captivated by the quaint culture and organic society townsfolk perceived in the country' (54).

If Jefferies' mainly non-fictional impressions of country life failed to satisfy late Victorian readers, then they could always turn to Thomas Hardy. His first Wessex novel, *Under the Greenwood Tree,* appeared in the year after Jefferies began writing. Like Jefferies, he was born deep in the Wessex countryside, but unlike Jefferies he remained there for most of his life. As Keith (1988: 86) has remarked,

> Hardy had personal experience both of traditional rural life and of the modern patterns that had taken its place, and in this respect he was in a unique position to portray the development of his own region through its most crucial period in history.

Hardy himself was at pains to stress that he was writing about rural life as it was. In his preface to the Wessex edition of his novels in 1921, he wrote, 'At the dates represented in the various narrations, things were like that in Wessex' (in Keith 1974). Victorian readers were readily drawn to his 'pastoral' novels and their portrayal of a surviving rural culture – *Under the Greenwood Tree, The Mayor of Casterbridge, The Trumpet Major* and *Far From the Madding Crowd,* but were shocked by the social commentary of his later novels like *Tess of the D'Urbervilles* and *Jude the Obscure* (Eustice 1979; Williams 1973). Even so, a review in *The Speaker* in 1891 reflected what attracted most readers to Hardy when it wrote of *Tess,* 'It deals with the old country, the old scenes, and, we might almost say, the old people' (in Lerner and Holmstrom 1968).

That Hardy became a cult in his own time had much to do with the late Victorian unease about industrialism. He offered a detailed picture of the people and customs of a rural world which was rapidly coming to an end. In this respect he reflected the growing interest in pre-industrial folk culture, led by William Morris and the arts and crafts movement. Yet, unlike Morris and his equally urban colleagues, Hardy wrote from the perspective of an authentic, if middle-class, countryman. He is therefore part of an enduring tradition of English country writing which has set out to record through direct experience and knowledge the remnants of traditional country life. Hardy's contemporary George Sturt, writing from the experience of a country craftsman, established for several decades the deserved reputation of an accurate observer of the changing country scene in his two most famous books, *Change in the Village* (1912) and *The Wheelwright's Shop* (1923) (published under the pseudonym George Bourne). Sturt eschewed the sentimental rural nostalgia of his times, yet regretted the passing of the old ways (Williams 1973). During the 1930s this combination of realism and regret set in the context of detailed countryside description became popular reading. H. J. Massingham's meticulous accounts of local history, crafts and customs gleaned from his years of pottering around the Chiltern countryside had a particularly large following (Keith 1974). His numerous books were complimented by a rising group of what Cavaliero calls 'farmer novelists' such as A. W. Freeman and A. G. Street, whose stories of farm folk provided first-hand, practical accounts of agricultural methods and traditions (Cavaliero 1977). Equally popular in the thirties were Adrian Bell, H. E. Bates and George Ewart Evans whose documentary-like stories, drawn from their long experience of living in the country, captured the sense of the eternity of the land and the hardships and pleasures of rural life. To these we must also add Flora Thompson's trilogy, *Lark Rise to Candleford* (1945), which reawakened turn-of-the-century country life, in Massingham's words, 'at the very moment when the rich, glowing life and culture of an immemorial design of living was passing from them' (Massingham 1945: v).

Most of these authors also wrote about the countryside from a non-fictional perspective, documenting agriculture, nature, landscape, rural traditions in local stories in newspaper columns, radio programmes and magazine articles. Arthur Street, in fact, became something of a radio personality, a sage countryman whose regular features on country matters and appearances on the long-running show *Twenty Questions* until well into the sixties made him a household name at least in southern England. An equally popular figure is Street's successor, Ralph Wightman, another farmer–writer, who since the fifties has established a national reputation with his books, country-diaries in *The Guardian* and radio and television appearances in which he covers every imaginable aspect of country life. Wightman is part of the resurgence of popular British interest in country matters that has occurred over the past twenty or so years. While a good deal of this has involved the re-publication of many of the old country writers like Jefferies, Bourne and Street, there is a new strain of literature, combined with radio and television programmes, dedicated to the documentation of country life. Some, like Richard Muir's study of the English village, are historical in their approach (Muir 1981). Others, like Robert Blyth's survey of a Sussex village, *Akenfield*, describe contemporary rural society (Blyth 1969).

Much of the country life writing which I have just described is underlain by a strong back-to-the-land sentiment. As the next chapter explains in more detail, the back-to-the-land movements which emerged in Britain and the USA in the late nineteenth and early twentieth centuries, were accompanied and encouraged by a variety of publications extolling the benefits of country and farm life to city dwellers. This perspective has been particularly dominant in North America, where most of the country life literature has been written from the perspective of urbanites who have retreated to the countryside. One of the earliest examples of this is Donald G. Mitchell. In numerous articles in *Harper's Magazine* (of which he was editor) and especially in a widely read book *My Farm of Edgewood*, first published in 1863 and re-issued in 1884 and 1891, Mitchell pioneered what was to become a century or so of literature by writers-turned-farmers. With a mixture of romanticism and practical wisdom, his book describes the business of taking up farming as a way of life, with hints on bee-keeping, cattle-rearing, milk and egg production and other farming activities.

This became the standard formula for the literature of the back-to-the-land and country life movements that surfaced at the turn of the century. The general advantages of commuter-agrarianism were promoted by the leader of the country life movement, Liberty Hyde Bailey and by Theodore Roosevelt. But it was the more down-to-earth and autobiographical advice of writers such as Bolton Hall, Edward Bok and Gustav Stickley that drew the greater readership. Hall's two books, *Three Acres and Liberty* (1907), and *A Little Land and a Living* (1908), together with the arts and crafts inspired ideas on simple living of Bok and Stickley were part of a general country life propaganda (Shi, 1985). Bok wrote in the *Ladies' Home Journal* and Stickley was founder, in 1901, of *The Craftsman: An Illustrated Magazine for the Simplification of Life*. These magazines and others like *Country Life* carried a growing number of articles on a wide range of aspects of country living. By the 1920s the country life movement had run its course, to be replaced by a more idealistic back-to-the-land philosophy led, as we see in the next chapter, by Ralph Borsodi. His 1933 book of practical advice based on his own experiment with subsistence homesteading attracted a good deal of attention especially in the depths of the depression years.

Although the country-living/back-to-the-land cult had died away by the 1930s, there remains a steady, if thin, trickle of books by urbanites who have taken up farming. Among the most well-known of these authors is Louis Bromfield, a writer-turned-farmer who, in the 1940s, published several books, both fictional and non-fictional, drawn from his experiences of his own farm at Malabar and its rural surroundings. The most famous are his novel *Pleasant Valley* (1945) and his journal of daily farm life *Malabar Farm* (1948). With the resurgence of back-to-the-land ideology since the 1960s, Bromfield has experienced something of a revival, but has also been joined by a new generation of books and magazines about farm and country life. Some, like Donald Hall's memories of boyhood summers on his grandfather's New Hampshire farm in *String Too Short to be Saved* (1961) and Howard Good's *Black Swamp Farm* (1967) – 'a faithful portrayal of farm life . . . as experienced and witnessed by one who was born in a log house on the farm there in the last century' – are nostalgically historical in flavour. To this heritage perspective we can add the proliferation of autobiographical accounts of latter-day back-to-the-landers like John Graves' 1980 book, *From a Limestone Ledge*, which is subtitled, 'Some essays and ruminations about country life in Texas', and Noel Perrin's four-volume description of the ups and downs of hobby-farming (Perrin 1978, 1980, 1984, 1991). A particularly good example of this genre, however, is Richard M. Ketchum's *Second Cutting* (1981), a collection of essays on country living by a 'sometime farmer' originally published in Ketchum's own magazine *Blair and Ketchum's Country Journal*. This magazine is one of several that have appeared in recent years which specialise in back-to-the-land/hobby-farming topics. Yet magazines like *Blair and Ketchum's* and its popular Canadian counterpart *Harrowsmith* have a large middle-class urban readership who clearly take pleasure in learning about gardening, composting, fruit preserving, goat-rearing, alternative energy projects and other aspects of country living.

CHILDHOOD PASTORALS

Like so many of our world images, our perceptions of the countryside are first acquired in childhood. These may come from the direct experiences of a rural upbringing or the more selective encounters of vacations and those interminable Sunday afternoon walks. They are perceptions too which, as children, we absorb from older generations and the values of the culture in which we are raised. Yet many of our most durable and stereotypical images of the countryside come from the literature of our childhood. Children's literature as such does not begin to appear until at least the mid-nineteenth century. To be sure, before this young readers were exposed to the prevailing images of the countryside drawn by the poets and novelists of the early nineteenth century, for these would certainly have been part of the normal upbringing of the educated classes. But it is only in the period from 1860 to 1890 that a significant body of literature specifically written for children emerges (Ellis 1968). Given the importance of rural and natural settings as well as the survival of romanticism in the general literature of the times, it is not surprising that nature and the countryside was from the outset a central theme for this new literary stream. Moreover the essential ingredients of children's literature, and, of course of the films and television programmes which are increasingly replacing it, are fantasy and adventure, for which the countryside is a perfect setting; a familiar yet at the same time imaginary world which serves as a retreat from adult-dominated everyday experience.

Among the first to recognise this was our old friend Richard Jefferies whose 1870s novels *Wood Magic* and *Bevis* follow the adventures of two boys in the woods and meadows of the English countryside. The use of natural settings as a backdrop for adventure stories, in fact, has produced some of the classics of British children's literature. This theme continues well into this century with Rudyard Kipling's *Puck of Pook's Hill*, Arthur Ransome's *Swallows and Amazons*, and the nature fantasies of J. R. R. Tolkien (*The Hobbit*) and C. S. Lewis, each in their different ways identifying childhood adventure with natural surroundings. In North America the turn-of-the-century nature movement spilled over into children's fiction. The great outdoors became the setting in which 'tenderfeet were skilfully guided to exchange the ways of urban adults for childhood in Arcadia' (Schmitt 1969: 124). This, in a proliferation of children's nature adventures between the 1900s and the 1920s, was the world of Gene Stratton Porter's *A Girl of the Limberlost* and Howard Gasis's *Canoe Boys and Campfires* and *Motor Maids at Sunrise Camp;* of Thornton Burgess's Boy Scout series and George Bird Grinnell's *Jack the Young Canoeman*; of Ernest Thompson Seton's Canadian adventures of *Rolf in the Woods* and *Two Little Savages* and of Lillian Garis's *Bobbsey Twins* (Schmitt 1969).

Of course, generations of children have been presented with a gothic view of a hostile and terrifying natural world in traditional fairy tales. Yet the growth of modern nature appreciation has offered them an alternative image, not only in the romanticised treatments described above but also in the nature-in-the-raw images of writers such as Jack London. His *White Fang* and *Call of the Wild*, first published at the beginning of the century, remain the classics of nature adventures which set out to paint a realistic picture of wildlife. This educational tone has become an important theme in modern children's literature, especially with the rise of environmental awareness. Nature books and magazines, often published by naturalist organisations, form a significant proportion of non-fictional material for children. Nature and wildlife too have become popular themes in children's television programmes and movies, most notably through the productions of the Disney studios.

In much of children's literature, however, especially that written for very young readers, the animals and the natural settings in which they live are quite different. They are the indigenous inhabitants of a sentimentalised countryside; anthropomorphised creatures, as Margaret Blount has so aptly observed, set in 'a familiar and charming world in which nothing really

64

Plate 2.4   Beatrix Potter's Peter Rabbit: setting the animal theme in children's
literature for generations to come

goes wrong' (Blount 1974: 135). For generations of young children animals
are the country folk – the real inhabitants of villages, woods, fields and
river banks, with adult human beings either nowhere in sight or at best
hostile figures from an alien world. Given the immense and enduring
popularity of her little books, Beatrix Potter is probably more responsible
than any other writer for imbuing children with this perception of nature

65

and the countryside. First published in the 1900, Potter's stories have sold millions around the world in the intervening years and show no signs of losing their popularity as one of the staples of young children's reading. They have been adapted to television, have even been made into a ballet and have generated a range of commercial spin-offs from ceramics to clothing.

To Beatrix Potter, however, we must add Kenneth Grahame's classic animal fantasy *Wind in the Willows* and, of course, A. A. Milne's *Winnie-the-Pooh*, who with his woodland friends is probably the most popular of all country figures in children's literature. In both books, of course, the animals are surrogate figures in a satirical exploration of human relationships which has entertained generations of adults. But for most children the animals, especially in *Pooh*, are just cute and funny. Like Christopher Robin they admit young readers to an intimate and affectionate relationship with nature. This has become pretty standard fare in young children's literature, from Alison Uttley's tales of Little Grey Rabbit and Little Red Fox to Sarah Cotton's series of Ladybird Books about the adventures of a mischievous rabbit called Tasseltip. Ladybird Books have led the way in this type of children's literature. Their 401 series, for example, which is aimed at 3 to 6 year olds, includes titles like *Bunnikin's Picnic Party, Piggly Plays Truant, Downy Duckling, Bob Bushtail's Adventure, Mr Badger to the Rescue*, and *The Bunney-Fluffs' Moving Day*. In these and in all the other books of this genre the animals live in quaint little cottages, surrounded by flower and vegetable gardens, and looked after by benevolent Mr and Mrs Rabbits, Squirrels, Ducks, Badgers and so on.

That it is a type of children's literature which is still predominantly British has much to do with the general national sentimentality towards the countryside. However, through aggressive marketing, the publishers of the Beatrix Potter and the Pooh books, as well as the Ladybird series have transferred this portrayal of the English countryside to young North American children. But North American writers and publishers too have cashed in on the popularity of animal themes. Harper and Row's *I Can Read* series uses predominantly animal characters, notably in Russell Hoban's Frances books about the adventures of a remarkably childlike badger. Alongside these books aimed at very young children, however, there exists a more complex anthropomorphised treatment of animals. The classic is E. B. White's *Charlotte's Web* which has enchanted children (as well as adults) for almost half a century with its portrayal of life, and especially rural life, as seen through the eyes of a wise old barnyard spider. Yet, while Charlotte is the focus of the story, the farm animals – notably the pig – with whom she chats and the kindly farm family who care for them, form a powerful pastoral image, which draws directly on the agrarian romanticism of North American culture. Popular, as we have already seen, as a setting for adult romantic literature, the farm and its rural surroundings have also long been a favourite place for North American children's stories. A common theme in the huge trade in younger children's picture books appears to be that of visiting grandparents' farms. Presumably intended to evoke the sense of returning to ancestral rural roots, these stories follow an unwaveringly sentimentalised formula which portrays farm life as a utopia of happy barnyard animals watched over by kindly old folk. For older children Kate Douglas Wiggin's *Rebecca of Sunnybrook Farm* and Lucy Maud Montgomery's *Anne of Green Gables*, are the lasting classics, still enthusiastically read eighty or so years after their original publication. *Anne of Green Gables*, set in turn-of-the-century rural Prince Edward Island, has become part of the Canadian cultural identity, re-issued in countless editions (including many foreign language editions), as well as dramatised for stage and screen.

Much of the abiding attraction of *Sunnybrook Farm* and *Green Gables* is that they portray an old-fashioned rural world, a kind of a magic yet recognisable kingdom which lies beyond the experience of modern life. The fact, too, that their market is primarily female reveals a profound gender bias in children's literature in which the countryside is the preferred setting for girl's stories. A particularly insidious example of this is the British writer, Joyce Lankester Brisley's *Milly-Molly-Mandy* series. First published in 1928, it was re-issued in 1972 by Penguin who told their young readers that 'Milly-Molly-Mandy and her Mother and Father, Uncle and Aunty, Grandpa and Grandma are just as busy as ever ... in their nice white thatched cottage in the country'. The children in these books spend their time in trips to the village shop, playing in the fields, helping Mummy in the kitchen and Daddy in the garden, feeding the farm animals and so on. The most popular of all girl's country books, however, must surely be the pony stories. Generations of girls have seen the countryside through the pages of the horse riding adventures of Christine and Diana Pullein-Thompson, Judith Berresford and Patricia Leitch.

This is not to say the countryside is the exclusive preserve of girls' stories. In English children's books in particular, it has long been the preferred setting for adventure stories aimed at both sexes. The Edwardian E. E. Nesbit's still-popular novels *The Railway Children* and *The Story of the Treasure Seekers* are set mainly in a country cottage and the surrounding countryside. This 'middle-class, Arcadian world' as Carpenter (1985) has described it, dominates the children's literature of the inter-war years with writers such as Richmal Crompton, Walter de la Mare and, of course, C. S. Lewis. And millions of mainly urban, state-school children have assimilated images of an idyllic countryside through the 'summer hols' adventures of boarding school children in the enormously popular *Famous Five* series by Enid Blyton. A traditionalist of the first order, Blyton portrays rural England very much in the 'Heart of the Country' spirit; a place where children can still feel secure within a modern version of the old social order yet at the same time roam free across an accessible countryside. Even amidst the hi-tech, sci-fi comic book, television and film-dominated world of Batman, Star Wars and the profoundly urban Ninja Turtles, the countryside seems to retain its appeal for children as a place for imaginary escape and adventure.

THE COUNTRYSIDE INDUSTRY

The development of countryside literature, and certainly of its adaptation to the electronic media, that we have explored in this chapter has a distinctly commercial edge to it. From quite early on, the publishing industry recognised the marketing opportunities of tapping the rural nostalgia of a largely urban readership. Films, radio, television, and ultimately consumer culture in general have taken this to its commercial extremes, transforming in the process the armchair image of the countryside into a profitable commodity.

Publishers have been cashing in on the countryside theme for at least two hundred years. The development of new printing technologies at the beginning of the nineteenth century opened up the book and magazine market on a scale which offered unprecedented commercial opportunities. Travel guides, illustrated nature anthologies and poetry collections, popular editions of the country house classics and of not-so-classic rural romances, magazines featuring articles on rural topics and others devoted entirely to the celebration of country life – all these formed a large sector of nineteenth-century publishing on both sides of the Atlantic. By the end of the century, as rural nostalgia and accessibility reached new levels, it was quite apparent that the countryside had become one of the most profitable themes for the publishing industry. In the USA, as we have seen, it tapped into the country life/back-to-the-land and nature sentiment of the times in a big way. In Britain it is perhaps best illustrated by the cult which grew up around Thomas Hardy. Not only did Hardy's novels attract a huge readership, published in several editions by the 1920s, but they were accompanied by a flood of handbooks and guides to 'Hardy country' (Cox 1976).

During the inter-war years the British publishing industry in particular made a significant contribution to the development of national countryside sentiment. Batsford fashioned its publishing house entirely around this theme, with its various series on the English countryside. Longman produced its English Heritage series, while other companies such as Victor Gollancz and Robert Hale specialised in publishing the work of country writers like H. E. Bates and A. G. Street. None of this, however, matches the degree of commercialisation of countryside literature of recent years. This has involved both the extensive re-issuing of earlier classics of country writing and the proliferation of modern publications on every imaginable countryside topic.

There appears to be a ready market for new editions of the old country writers. Between them four publishers – Oxford University Press and three smaller companies, Quartet, J. Clare Books and Horizon Handmade Books – have re-issued Richard Jefferies' major works. In its Country Library Publications series, Penguin has re-published several of the classics, including George Bourne's *Change in the Village*. Cobbett's *Rural Rides* and *Cottage Economy* are now each available in three editions including a lavishly illustrated 1984 publication by Webb and Brown. A 1978 edition of W. H. Hudson's *A Shepherd's Life* with specially commissioned engravings by Reynolds Stone and publicised as a 'faithful record of bygone Wiltshire days', typifies the marketing of this type of publication, as does Batsford's re-issue of its 1930s village series in newly illustrated editions. Flora Thompson's *Lark Rise* trilogy has been subjected to an inappropriately sentimentalised treatment in a recent edition by Century Publications bound in floral designs and illustrated with depictions of country scenes painted in the Victorian sentimental style. Along similar but considerably more commercialised lines there is Webb and Bower's resurrection of Edith Holden's *Country Diary of an Edwardian Lady*, which has become something of a cult in its own right. Finally among these and many other examples too numerous to mention, there is the revival of the Hardy industry. The Wessex novels are available in several editions, some of them freshly illustrated and newly packaged. But it is the memorabilia and the nostalgic celebrations of the Wessex landscape which reveal the true extent of the Hardy industry. Illustrated descriptions of Hardy country abound, among the best examples of which are the naturalist Gordon Benningfield's water-colour illustrated *Hardy Country*, published by Penguin, and Desmond Hawkins' delightful juxtaposition of excerpts from Hardy's novels and the landscapes in which they are set.

The modern 'Hardy country' books are part of a general wave of recent popular publications about the British countryside. A large proportion of these serve the 'coffee-table' trade – lavishly illustrated volumes, aimed mainly at the tourist and Christmas gift market. Bookstores devote whole sections to this type of literature. Most of the large publishing houses have tapped into this market, while a few like Weidenfeld and Nicholson, Robert Hale, again Batsford and several small presses have become specialists in the trade. *Country Life* magazine too has capitalised on its upscale image with the publication of its *Country Life Books* of various regions of rural Britain. At their simplest level these are landscape descriptions – the countryside as a series of picturesque and peaceful scenes – among the best examples of which are the pictorial volumes which have been produced to cash in on the popularity of country writers. One of the best-sellers of this type is *James Herriot's Yorkshire*, the author's own pictorial celebration of the Yorkshire Dales. In addition to these there are numerous picture books which present the standard images of the English countryside – farmscapes, villages, thatched cottages, country inns and churches, castles and stately homes. Partly aimed at an overseas tourist market, they are just as likely to be found in North American as in British bookstores.

Publishers, however, have recognised a growing demand for more detailed descriptions of the countryside. The common formula of many guide books, in fact, is a combination of informative narrative and glossy illustration. Among the best examples of these are countryside volumes sponsored by Shell, the petroleum giant. In addition to its bestselling *Guide to the Countryside*, Shell has supported over the past decade or so a number of books packed with beautifully illustrated information on different aspects of the countryside. The commercial opportunities of linking tourism and the countryside ideal have not escaped other companies; BP has also sponsored similar publications. The Automobile Association, too, has produced its own versions of countryside books, designed presumably to attract more drivers, and therefore more members, on to Britain's already congested country roads. Despite their tourist orientation, however, the Shell and AA books satisfy a significant sedentary interest in the countryside. Much of their popularity stems from their strongly heritage theme. In this respect they are part of a growing appetite for well-illustrated books which tap rural nostalgia. A particularly profitable theme seems to be popular histories of the countryside, like Hodder and Stoughton's county by county *Making of the English Landscape* series, edited by W. G. Hoskins along the same lines as his original book with the same title. Historians, both amateur and professional, make a good living out of this type of publication. Among the most productive is Richard Muir, who has authored or co-authored at least sixteen illustrated volumes on various aspects of the English countryside. Through its many publications, the National Trust, of course, has probably done as much as any commercial publisher to foster public interest in and financial support for both the natural and cultural heritage of Britain.

Nostalgic interest in the countryside, however, has fuelled a vigorous trade in less intellectual treatments of rural heritage. For many consumers of coffee-table publications, mainly pictorial volumes depicting traditional rural architecture and customs seem to suffice. Barns, farmhouses, cottages, old farming methods, pictures of Victorian village life and so on have become standard fare for the nostalgia industry. The portrayal of the countryside through images of old buildings is particularly apparent in North American publishing, which, although largely disinterested in general countryside books, has been quick to tap the prevailing view that rural heritage is largely a matter of architecture. Barns and farmhouses are especially popular subjects.

That country themes have become a profitable business for the publishing industry is evident in more than just the direct celebrations of rural life and landscape that I have just described. It extends to a whole range of topics which have become part of the very stuff of post-industrial consumer society. A quick perusal of the latest issue of *Books in Print* reveals three pages of titles on 'country' topics ranging from antiques to wine-making. Country cookery is a particularly marketable subject, with book after book on traditional country recipes, bottling, preserving, wine-making and all the 'natural' kitchen activities associated with country life. But these are simply part of a larger trade in the marketing of country lifestyles. Country recipes must be cooked in country kitchens, which must be decorated with country wallpaper and furnished with country antiques, and located in a country home and garden. And if those attracted to this fabricated lifestyle, in practice or day-dreams, do not have the time or inclination to read one of the many books on the subject which line the shelves of their local bookstore, then they only have to turn to the pages of magazines.

There has been a boom in country life magazines in the past decade or so. Some, like the long-running British magazines *Country Life* and *The Countryman* and the more recent North American back-to-the-land publications such as *Harrowsmith*, *Blair and Ketchum's Country Journal* and the *Mother Earth News*, set out to present a serious treatment of country life. But, with the exception of *The Countryman*, they are suffused with country lifestyle promotion. While, as we see in the next chapter, this reveals much about the values of those who have acquired country properties, it also suggests that there is a healthy urban market for the consumerist trappings of country living. This is even more apparent in the 'home and garden' magazine trade, which has adopted country-style as one of the defining motifs of domestic design. The American monthly, *Country Living*, published by the Hearst Corporation, is a standard item on the magazine racks at the supermarket check-out counter. Aimed mainly at a female readership (it also has an all-female editorial executive), it portrays and purveys in articles and advertisements alike a lifestyle soaked in contrived rural nostalgia. William Morris prints, pine furniture, dried flowers, wicker baskets, patchwork quilts and earthenware pottery abound, all set in country-cottage surroundings. This formula is repeated in *Country Living's* unrelated British namesake, although with a predictably more gentrified veneer. It is also a regular feature of the cookery and home fashion sections of women's magazines in general, which, in a long line stretching back to *The Ladies' Home Journal* and the *Ideal Home* magazine continue to peddle the thoroughly suburban association of domesticity and country-style living.

Integral to the commercialisation of this ideal is the production of the consumer goods which give it tangible expression. The popularity of the countryside motif in interior design has spawned a number of

Plate 2.5 'It was a Simpler, Gentler Time and Place': nostalgia in ornamental china

specialised industries, including the manufacture of replica antiques, soft furnishings in traditonal designs and materials (Laura Ashley built an

72

Plate 2.6 The marketing of pastoral purity: Horlicks' advertisement, 1901

international corporate empire out of this), kitchen-ware and, as Plate 2.5 shows, ornamental china. Even organisations like Britain's National Trust have cashed in on this market by commissioning a range of household items, from tea-towels to table-mats, designed around nature and country-side themes, which it sells by mail order, in its own shops and in the classier department stores. And if these fail to satisfy those wishing to surround themselves with countryside images they can always bring them home from the many retail outlets which, as we shall see in the next two chapters, have sprung up in the countryside itself precisely to exploit this market.

No discussion of the commercial exploitation of countryside sentiment would be complete without reference to its role in advertising. Since the emergence of advertising as an industry towards the end of the last century, nature and country life have been used to promote consumer goods. Adver-tising has become a powerful, perhaps *the* most powerful, reflector and manipulator of values. It relies heavily on the tapping of deeply held ideals and popular images in often subliminal associations, and the moulding of these into new and simpler myths. The countryside thus offers a perfect set of images for the promotion of certain products. The food and drink industry has long had a special preference for this approach, as can be seen in the 1901 Horlick's advertisement in Plate 2.6, in which the idea of the product's purity is projected with a virginal milkmaid and a contented cow set in a pastoral scene. These nostalgic connections between food, the land and wholesomeness remain popular among food companies today and are typical of what Marchand (1985) has called the 'visual cliches' of advertising.

Idyllic natural settings and rural landscapes are particularly popular advertising devices, used to convey a sense of naturalness and purity, not only to food, but also to a variety of everyday products (Plate 2.7). The association of nature, fresh air and open space with the enjoyment of a cigarette has long been a standard advertising formula. This is epitomised by the globally famous Marlboro Man advertisements in which the figure of the cowboy on the open range evokes a sense of healthy manliness and freedom which obscures the grim realities of nicotine addiction.

That the armchair countryside has been integrated into the modern commercial world is a measure of the extent to which cultural values in general have been reduced to trivialised images. In this respect the country-side for the vast majority who have no direct experience of it, may be no more than a series of distorted impressions of nature and country life which only serve to perpetuate the mythology of the countryside ideal. Yet beneath this commercial veneer lies the more serious interpretation of both the ideal and of the countryside itself. As we have seen, this runs in several well-worn literary and artistic threads which have woven some

Plate 2.7 Idyllic natural settings are used to market a variety of everyday products: 1987 advertisement

deeply held beliefs about the virtues of rural life and landscape into the very fabric of our culture. That this has been dominated by a selectively idyllic view of the countryside has guaranteed its idealisation across many generations and strongly influenced the use and treatment of the country-side itself.

3

A PLACE IN THE COUNTRY

The countryside ideal could not have been sustained by abstract ideology and imagery alone. Reinforcing, and in many ways closely linked with the philosophical musings on the virtues of the country over the city and the armchair sentiment for country life and landscape, has been a growing movement to seek escape in the countryside itself. From this has emerged the perception and use of the countryside as an amenity; an environment set aside for urban pleasure and relief. This has played a powerful role both in the nurturing of the countryside ideal and in the transformation of large areas of rural landscape to conform to this ideal. At its most popular level, it has converted the countryside into a mass recreational amenity; an extended playground and vacationland for society at large. This forms the subject of the next Chapter. This Chapter, however, explores the use of the countryside as a more exclusive and personal amenity, one which is driven by the desire for a more permanent retreat; for, in other words, a place in the country.

The country retreat takes a variety of forms. For a long time, as we saw in Chapter 1, it was the preserve of the élite and the wealthy who brandished their status with grand houses and landscaped parks. However, with the rise of an affluent and increasingly mobile urban middle class, they have been joined by the somewhat more modest properties of exurban commuters and country weekenders, by the small-holdings of the hobby-farming fraternity and by those who have set up a variety of businesses, from country inns to antique shops, to exploit the amenity use of the countryside. And, finally, in a quite different form of retreat, we can add the various manifestations of the back-to-the-land movement, from the utopian agrarian communities of the nineteenth century to the rural communes and alternative-lifestyle seekers of today.

THE COUNTRY HOUSE

There is historical logic in beginning with the country house, for this is where the trend for using the countryside for recreation and retreat

77

originates. It has its roots in Republican Rome, when it became fashionable for the very rich to own not only a house surrounded by a park in Rome but also a country house within easy reach of the city (a *villa suburbana*). An expensive seaside villa and even another, smaller country house would also have been part of the lifestyle of the exceptionally affluent (Balsdon 1969). The country estate and the *villa rustica* became an integral part of late Roman culture, symbolic of an increasingly urban perception of the immediate countryside as an amenity for the leisured classes. In fact, the word amenity derives from *amoenitas*, a Roman term for the aesthetic and sensory pleasures of a country retreat (White 1977). Country villas were invariably palatial and surrounded by ornamental gardens with swimming pools and fish ponds as well as by estates with woods and working farms. They offered, therefore, the oppportunity for both idle relaxation and the pursuit of more active rural pleasures. 'Whenever I am worn out with anxiety, and want to sleep,' wrote the poet Martial, 'I go to my farm.' Cicero, too, was able to 'dwell at length on the many delights of country pursuits'. These included hunting, walking in gardens and woods, enjoying the scenery and acting out the role of a gentleman farmer (in White 1977).

It is this ideal which provides the link between the Roman country villa and the country house of modern civilisation. In the late thirteenth century, the classical description of life in the villa as an aesthetic retreat for men of letters was the inspiration for early Italian country houses. And, in turn, it was the grand and formal style of the Renaissance country villa, with its ornate gardens and prominent settings, which influenced the English country estate in the sixteenth century (Newton 1971). The spread of this fashion to Elizabethan England was as much motivated by the attractions of property investment as by the desire for a country retreat. Indeed, as we have already seen in Chapter 1, social status and the accumulation of wealth were the driving forces behind the proliferation of country estates in the late sixteenth century. As an integral component of national and political life these estates were, in most cases, the principal residence of their owners, and therefore the focus of everyday life for a predominantly country-based nobility and gentry. However, its Arcadian literary associations are ample testimony to the significance of the early English country estate as a place of leisure, relaxation and the enjoyment of nature and scenery. The country estate was especially valued as a recreational amenity, a fact which is apparent in the deer-parks which were laid out and in the substantial body of writing which describes the pleasures of rural sports on the gentleman's estate.

Although wealth and status continued to be the driving forces behind the great wave of country estate development in the eighteenth century, it was accompanied by a growing appreciation of its value as a retreat for the leisured classes. There were several reasons for this. Firstly, the emergence of the new class of landed gentry produced by the growth of mercantile capital introduced a greater sense of differentiation between the city as a place of business and the country house as a place of relaxation. It led also to greater class distinctions in the countryside with the rise of what Oliver Goldsmith called 'polite society', which used the country house for leisurely pursuits increasingly apart from the rural population which was the source of much of its income. Another factor by the 1720s was improved transportation: better carriages, fast phaetons and improved roads which enabled polite society to pay visits from one country house to another (Girouard 1978). The country house became a focus for fashionable folk; a place for socialising and uninterrupted leisure. 'The English,' to quote Hugh Prince (1967), 'wanted space in which to take vigorous exercise, to canter a horse or to survey the scenery' (6).

This was reflected in the designs of new and reconstructed houses, but was expressed most clearly in a growing interest in the landscapes which surrounded them. While the work of the great landscape gardeners was part of the general spirit of improvement which transformed the English countryside, it was also very much directed at the use of the country estate as a recreational amenity. The replacement of the rigid formality of the grounds of country houses by the picturesque layouts of Charles Bridgeman and William Kent stressed the informal enjoyment of nature and scenery. Bridgeman achieved lasting fame for his introduction of the 'haha' (originally a French invention from the beginning of the seventeenth century), a ditch which kept animals out of the grounds while preserving an uninterrupted view of the surrounding landscape (Jellicoe and Jellicoe 1975). The essence of this design was the idea of gardens as landscape paintings, with carefully planned 'natural' perspectives (Hyams 1971). In his designs for Rousham and Stowe, Kent went one step further towards the creation of a recreational parkland. At Stowe, he introduced the garden-circuit, a layout of driveways and footpaths, which permitted leisurely walks around the extensive park. Points of scenic, botanical and architectural interest were carefully incorporated into this design (Girouard 1978).

With the work of Kent's successor Lancelot 'Capability' Brown the idea of the country estate as a leisure park was extended to large chunks of the English countryside. With his completion of Stowe, his most famous work at Blenheim and many other projects, Brown became the 'darling of the fashionable world' (Newton 1971). His landscape compositions, with their broad sweeps of grass, encircling woodlands, clumps of trees, artificial lakes and altered stream courses were aimed with aesthetic purpose at the perfection and framing of the prospect from the house (Clemenson 1982). Newton has suggested that Brown may not have been aware of the idea of designing outdoor space for human use (Newton 1971), but his work was certainly reflective of the growing attraction in the eighteenth century of the country house as a pastoral as well as a fashionable retreat.

By the end of the century, with the appearance of the first self-professed

Plate 3.1 The English country house: Wilton, Wiltshire

landscape gardener, Humphrey Repton, the fashion for laying out the grounds of the country house for the maximum of visual effect and physical enjoyment had become firmly established. Repton, the great compromiser in the controversy surrounding the criticism levelled at Brown's work by the exponents of the new version of the picturesque, combined formal and romantic traditions, to produce a setting which, in his own words would 'call forth the charms of natural landscape' (in Newton 1971). Repton introduced what was to become the nineteenth-century convention of a house surrounded by a beautiful garden which led to a picturesque park. By 1814 he was famous enough to be referred to in Jane Austen's novel *Mansfield Park*, as the man to employ to create grounds which would be 'the admiration of all the country'.

Repton's fame rested not only on the good taste of his landscaping but also on his ability to create scenes which could be shown off to fashionable society. He worked during a period of country house living which has been described as an age of informality, during which the rising distaste for urban conditions was expressed by the upper classes in a growing enthusiasm for spending six months or so in the country, enjoying nature and country pursuits (Girouard 1978). The layout of grounds became increasingly elaborate and ostentatious and the 'improvement' of the country property became a maxim for fashionable society. However, unlike the spirit of agrarian improvement earlier in the century, the emphasis was on the aesthetics of landscape and the creation of appropriate settings for the entertainment and leisure of 'society'. This, in Girouard's words was the 'golden age of the country house', in which recreation in the countryside in the Victorian grand manner was the height of fashion (Girouard 1978). Weekend parties and the seasonal hunting and shooting weeks to which all the right people would be invited were facilitated by the great improvements in transportation occasioned by the development of turnpikes and the arrival of the railway. Despite the opposition of some landowners to the railway, many were not averse to negotiating for a halt just beyond the gates of their estates. Indeed, the railway heralded a new era for the country house, enabling it to be used as a weekend retreat, or even as a residence from which to commute to town on a daily basis.

A significant factor in the evolution of the country house into a recreational amenity, however, was the investment in country properties of the 'new' money of the rising urban middle class. A place in the country for relaxation and entertainment became the fashionable way to spend surplus capital for the growing number who sought to join the ranks of the gentry. Alongside the estates of the old nobility and squirearchy sprang up the country properties of the industrialists, bankers, entrepreneurs and professionals that were becoming an increasingly influential element of the Victorian establishment. A combination of factors, beginning with the agricultural depression of the 1870s and 1980s, continuing with the introduction of estate duty in 1894 and culminating in Lloyd George's legislative attack on large landowners in 1910, led to the decline of the landed estate as an income-generating property. For the *nouveau riche*, however, this served only to increase the opportunities to acquire a grand place in the country. The sale of houses, parcels of estate land and, in some cases, entire estates supplied a growing demand for desirable country properties (Clemenson 1982). The launching of *Country Life* magazine in 1887, which to a large extent depended upon the advertising of such properties for its survival, illustrated the importance of a country house to an increasingly affluent upper middle class. *Country Life* contributed also to the growing eulogisation of the lifestyle of the English country gentleman (Girouard 1978).

The heyday of this ideal in British society was between 1900 and the late 1920s. Few new grand country houses were built but, for architects such as Lutyens, Voysey and Baillie Scott there was in the early decades of this century a steady demand for more modest country places. These and the older houses which survived continued, particularly in the Edwardian years, to be the focus of much of the social whirl of the affluent, who enjoyed a lifestyle of hunting, shooting and lavish weekend parties without being dependent on agricultural income for its support (Clemenson 1982). The increased mobility provided by the automobile played a central role in this, and, for the next twenty years, the country house and its escapist lifestyle remained a symbol of élite society.

In the space of a couple of centuries or so the nobility and gentry transformed much of the English (as well as the Welsh, Scottish and Irish) countryside into a landscape which, as Beresford (1966) has observed, is a product of leisure to a degree unmatched elsewhere. While the French landowners left the countryside to the peasantry, while the knights of Prussia were still feudal barons and while the pioneers of the New World were preoccupied with carving a living out of the wilderness, the 16,000 owners of the British countryside had the time to plant their gardens and their playgrounds at their leisure.

In a rural society founded largely on the principles of egalitarianism and the edicts of agrarian progress, there was little room in North America for the development of the leisured landscape of the country estate. The notable exception was the plantation culture of the South. The principal objective of plantation owners, of course, was that of profiting from an economy based on slavery and the staples trade rather than the creation of a country retreat. Yet the ante-bellum southern mansion was the focus for leisure and socialising for a ruling class which imitated many of the pretensions of the European gentry. With the abolition of slavery and its replacement by more arm's length exploitation in sharecropping and tenant farming, the social aspects of the southern mansion became more visible.

This established a standard of genteel southern living, which became increasingly anachronistic in the social context of the rest of the USA.

Further north, in the early seventeenth century, attempts were made to imitate the English country estate with patroonships along the Hudson river; vast acreages stretching up to sixteen miles along the shoreline, occupied by impressive manor houses (Newton 1971). However, it is with the rise of industrial and merchant capital in the north-east during the eighteenth century that the country house as a prestigious retreat first appears. Merchants and mill-owners in New England began building mansions away from the towns, and, later in the century, the growing influence of the romantic movement and of the picturesque ideals of the English landscape gardeners stimulated the laying out of grand country estates (Newton 1971). The most famous of these, George Washington's house at Mount Vernon and Jefferson's pastoral retreat at Monticello, are today national monuments and leading tourist attractions. Jefferson took a special interest in English gardens and was a great admirer of the work of Capability Brown. What Jefferson saw on a tour of English estates, including Stowe and Blenheim, profoundly influenced his design for Monticello (Hyams 1971).

Throughout the nineteenth century the wealthy citizens of the established cities of the north-east, in particular of Boston, commissioned the construction of houses in the country. The old manor estates along the Hudson River became summer homes as well as year-round country residences (Newton 1971). Architects established their fame and fortune with their designs for country properties. The most famous of these was Andrew Jackson Downing. Born in 1815, he had already published in 1841 his influential *Treatise on the Theory and Practice of Landscape Gardening Adapted to North America*, a work strongly influenced by the English 'gardenesque' school of landscape design (Hugo-Brunt 1967). Downing actively promoted the value of the country retreat, and although, as we shall see, was more interested in designing modest country homes for the middle classes, nevertheless was involved in the design of a number of grander country properties (Newton 1971). The heyday of the American country house was between 1880 and 1930, a period which saw the building of extravagant estates for wealthy tycoons (Aslet 1990). One of the most famous was the Biltmore Estate designed for George W. Vanderbilt in 1894 by Olmsted and Hunt. Today, as we can see in Plate 3.2, the house and its grounds still make an unequivocal visual statement about the wealth of old industrialism. Long Island was particularly popular with the country estate set. Scores of country places in a variety of styles from Elizabethan to neo-Gothic by architects like the Olmsted brothers, Bryant and Fleming, and Jens Jensen turned the eastern end of Long Island into a mecca for the wealthy (Newton 1971).

The grand country house tradition had reached its zenith in both Britain

Plate 3.2 'Extravagant estates for wealthy tycoons': the Biltmore Estate, North Carolina

and America by the 1920s. New houses have been built since but this has been a diminishing phenomenon, limited to the often eccentric activities of a declining number of the very wealthy. Tax reforms and, of course, the Great Crash of 1929 played a significant role in this. In Britain in 1980, only half of the houses which existed a century earlier were still used as a residence (Clemenson 1982). Many have been converted to institutional use, but, ironically, many have also become attractions for recreation and tourism, from historic properties to safari parks! Despite these changes, however, over 80 per cent of estates in Britain are still held by the great landowners (Clemenson 1982). Furthermore, in recent years there appears to have been somewhat of a revival of demand for a classy place in the country. A perusal of the property pages of any recent edition of *Country Life* for example will reveal an active market in prestigious country property. On the other side of the Atlantic the public brandishment of extreme affluence has also produced a new generation of lavish country retreats.

SUMMER HOMES AND WEEKEND RETREATS

As the country house has evolved into a place devoted primarily to the pursuit of leisure, the tendency to occupy it on a seasonal or occasional basis has increased. Even in sixteenth- and seventeenth-century England, while the majority of the gentry treated country houses as principal residences from which brief forays would be made into the city, there were those who regarded them as a country retreat for the summer and the game seasons. By the early nineteenth century the fashion for going up to London for 'the season' was beginning to be replaced by the counter movement of living principally in town and decamping to the country house for the summer (Coppock 1977). In North America, the country house often began life as a summer residence, only to be occupied year-round when transportation improved. In his design book for country homes, Woodward (1865) described the fashion for 'hastening gladly to these rural scenes with the opening of summer'(9).

The late nineteenth century, in fact, marks a turning-point in the history of the place in the country. From this point on the fortunes of the grand country house begin to decline. Its position at the centre of the rural residential leisure experience is gradually assumed by more temporary retreats. Variously termed vacation homes, cottages or (in modern academic parlance) 'second-homes', these sanctuaries from urban living, unlike their country house ancestors, have become a broadly based and eclectic middle-class phenomenon. Its ideological origins again can be traced as far back as Roman times, when the idea of a summer retreat in the countryside first appeared, as well as to the values which sustained the country house tradition in modern history. Yet its tangible origins, like so much of the contemporary use of the countryside, are to be found in the late Victorian period. Even before this the cult of the picturesque and the new attitudes to nature had initiated a vogue, particularly amongst writers, artists and intellectuals, for summer retreats well away from urban scenes.

By the 1880s the fashion had spread to affluent society in general. Leisure was becoming an increasingly important aspect of middle-class lifestyle, nostalgia for the countryside had reached new heights and an expanding rail network made remoter areas accessible. In Britain, a plentiful supply of farm cottages and other rural dwellings enabled a growing number of the middle class to satisfy their dream of a little place in the country without sacrificing their urban residences (Marsh 1984). The Edwardian era saw a vigorous trade in architect-designed country retreats, such that, by 1912, one of the leading purveyors of country cottage designs claimed that 'the weekend habit had become widespread, particularly around London', and that it was 'very much the thing to do to have a cottage in the country' (Elder-Duncan 1912: 11).

In North America the large areas of unsettled land, most of which were just the kind of wilderness environments which satisfied the growing passion for nature-based outdoor recreation, encouraged the construction of summer cottages and chalets. One such area was the Muskoka region of Ontario. Muskoka is a land of lakes and forests, perfect (in North American eyes at least) for a country retreat. In the 1890s, rail service from Toronto and a network of steamship services, together with the availability of largely free Crown land, saw the beginning of summer cottage development which has lasted to this day. Not all of those turn-of-the-century summer places in Muskoka, however, were simple cottages, for this was still largely a preserve of the affluent for whom a 'cottage' often meant a substantial house, preferably on one's own lake or island. This, of course, was very much in the tradition of the summer mansions of Newport and Long Island. Well into the 1920s, in fact, the summer retreat was most likely to be associated with the social tone of the places described by F. Scott Fitzgerald in *The Great Gatsby*:

> There was music from my neighbor's house through the summer nights. In his blue gardens men and girls came and went like moths among the whisperings and the champagne and the stars . . . On week-ends his Rolls-Royce became an omnibus, bearing parties to and from the city . . . while his station wagon scampered like a brisk yellow bug to meet all trains.
>
> (Fitzgerald 1925: 40).

Although the twenties were the heyday of summer residences for the affluent they also marked the beginning of what was to become a mass-movement. The key factors in this were the growth of paid leisure time (of which I shall say more in the next chapter), and the automobile. In many ways it was the automobile which provided the main impetus for the acquisition of more modest country retreats by somewhat less affluent urbanites. It had an immediate impact on countryside recreation and made it possible for families to travel freely between vacation or weekend homes and their urban residences. Its influence came early in the USA, so that by the 1920s areas like the Maine coast, the hills of Vermont and much of the coastline of the Great Lakes were dotted with summer cottages. In Britain, widespread car-ownership is a more recent phenomenon, but by the 1930s there were enough people with cars to make areas like the Lake District, North Wales and the south-west peninsula popular places for a holiday home.

The real boom in vacation homes, however, like that of recreation and tourism in general, has occurred in the last thirty years. Perhaps one should be cautious of referring to it as a mass-movement, for it involves no more than about 10 per cent of the population. Yet that translates into several millions of vacation home owners; it is clearly no longer the preserve of the very wealthy. In the USA the post-war growth of vacation homes has been phenomenal, particularly during the 1960s and 1970s. By the mid-seventies, it was estimated that three million American families, roughly 5 per cent, owned vacation homes (Ragatz 1977). This figure did not include the occupation of rental units, mobile homes, houseboats and old farm-houses which could substantially increase the total occupancy of second homes. On a per capita basis the vacation home is even more important in Canada. By the late seventies there were more than 250,000 cottages in Ontario alone with a growth rate estimated at 10,000 per annum (Priddle and Kreutzwizer 1977). In Britain, second-home ownership lags behind that of several other European countries, notably France (which has roughly three times the number of second homes per capita than the USA) and the Scandinavian countries. Nevertheless, the latest figures suggest that there are 200,000 second homes in Britain (Robinson 1990).

Because of differences in the definition of second homes and in methods of enumeration these data are not strictly comparable. For the most part, they are also fairly crude estimates. Furthermore, the ownership of a second home can encompass situations that have nothing to do with rec-reation. Nor do the data generally refer specifically to second homes in the countryside. However, amongst the scholars that have used and gener-ated these data, the assumption is made that most second homes are vacation and weekend retreats in non-urban locations. Of course the main attraction of a recreational property is to be able to use it as frequently as possible. Travelling to the cottage on weekends, especially in North America, has become just as important as longer stays during the summer. Proximity to the city, as with the country place phenomenon in general, therefore has long been a prime consideration for second-home owners. However, the spread of permanent residential development into the urban fringe has reduced its attractiveness for vacation homes. Moreover, in an era of express highways people now seem willing to travel fairly long distances to find the perfect retreat. Of course, in motorway Britain this means that virtually all of the countryside is now second-home territory. And in North America, it is surprising how far people are prepared to drive to their weekend cottages.

The most sought-after recreational properties are those with access to water, be it ocean, lake or river. This does not necessarily reveal any particular search for countryside, or burning desire to savour the delights of nature and landscape. The second home often may be more of a status symbol for an affluent, mobile and increasingly leisured society. Wolfe (1965) has advanced the rather opinionated argument that 'summering at the cottage' is a symbolic act:

> It is not for amusement; it is not an escape from the city, though that is what it most often seems to be . . . it is not even primarily for recreation . . . Today it symbolizes . . . the sense of belonging that all of us feel the need to demonstrate in one way or another.
>
> (Wolfe 1965: 7)

That it has become one of the conventions of middle-class society, does not, of course, diminish the second home's role as a recreational escape from the metropolitan environment. Nor does it eliminate the possibility that some people have acquired second homes in order to enjoy the countryside. In North America the traditional sentiment for natural environments in their own right, quite apart from their recreational resource value, has prompted urbanites to seek out wooded and mountainous areas for wilderness-like retreats. By the end of the nineteenth century thousands of Americans were experimenting with wilderness living in rustic cabins in the Adirondacks, Maine woods and even the Rockies (Schmitt 1969). This penchant for remoteness, for landscapes at the margins of settlement, also underlies the spread of second-homes into the Celtic fringes of Britain – North Wales and the Scottish Highlands in particular – and into the somewhat more domesticated settings of the Lake District and the Pennines. Furthermore, there seems also to be an undercurrent of rural nostalgia, of that attachment to the idea of a pre-industrial pastoral society, in second-home development. In Britain, in particular, pretty villages and the open but settled countryside have long been as popular for vacation and weekend retreats as for permanent country places. In the USA, too, there is a vigorous market in rural land for vacation properties, particularly in more marginal agricultural areas. Vermont and several other parts of New England are traditional vacation home areas which are especially valued for their pastoral landscapes as well as their more active recreation-oriented environments.

There is one fundamental difference, however, between vacation and weekend homes in Britain and North America which merits special attention. In Britain, as in much of the rest of western Europe, the neglible supply of unoccupied land, the high rural and exurban population densities and the generally tight controls on the development of recreational properties particularly in open countryside, have combined severely to limit new second-home construction. Although it could be argued that the British prefer old rural buildings for their vacation homes anyway, circumstances have ensured that, except in some urban seaside areas, old rural buildings are what they get. Despite strong competition from those seeking year-round country places, there continues to be a healthy market for second homes provided out of the existing rural housing stock, as well as for conversions of farm and other traditional rural buildings. Of course, a tastefully renovated cottage or a converted barn accords perfectly with the rural nostalgia which influences much of the demand for vacation homes in the countryside.

To some extent the North American vacation home market is also driven by a demand for traditional country dwellings, and, in declining agricultural areas, has been generally satisfied by a good supply of buildings for renovation and conversion. Yet most vacation homes are purpose-built. Because of the plentiful supply of land and the relative lack of controls on property development, both Canadians and Americans have considerably more freedom than the British in the building of their country retreats. Eclecticism in the structure and style of cottages is the result. Indeed, the 'cottage' may be anything from a humble trailer (caravan) or a tar-paper shack to a substantial, architect-designed house. However, running through cottage design there is a strong element of rusticity. This is particularly evident in the trend towards prefabricated cottages, in which log construction, the use of woods like cedar and pine, and an emphasis on simplicity of appearance as well as assembly are evocative of the sense of backwoods pioneering and closeness to nature in which cottage owners often indulge (Plate 3.3).

EXURBIA

It was Spectorsky (1955) who first coined the term 'exurb' to describe that great wedge of country residential development which stretches north and west for upwards of 100 kilometres from the edge of New York City. In Westchester and Fairfield Counties, along the North Shore of Long Island, up the Connecticut and Hudson Valleys are the landscapes of the latter-day country gentry; the commuter and retirement lands of Manhattan executives and professionals fulfilling their desire for status and retreat on their few acres of countryside or in their sprawling rural subdivisions. This kind of residential development has come to dominate the rural landscapes of most metropolitan regions in the western world. With it has come the broadening of the country place ideal beyond its élitist roots, and the

Plate 3.3   The purpose-built rustic escape: 'cottage-country', Ontario

redefinition of the character – social and economic, as well as visual – of the metropolitan countryside.

Although the building of turnpike roads and the development of improved forms of private carriages led to a trickle of middle-class migration to country villas within commuting distance of cities on both sides of the Atlantic in the late eighteenth and early nineteenth centuries, exurbia is overwhelmingly a product of the railway and automobile ages. With the arrival of the railway commuting for both business and pleasure became, for the first time, a practical proposition for anyone with the money and the inclination to take up residence in the country. For the social élite of east coast American cities it was an ideal popularised by a steady stream of books and magazine articles, as we saw in the previous chapter. Architects rushed to satisfy the demand for rural properties. 'The remedy for urban expansion and associated ills,' wrote George Woodward in his pattern book of country homes in 1865, 'is to go into the country.' Country properties, cheaper to buy and to manage than those in the city were available 'to people of modest means within an hour of City Hall' (Woodward 1865: 12). For the immensely popular rural architect, Andrew Jackson Downing, it was in rural residences that true taste was to be found. 'When smiling lawns and tasteful cottages,' proclaimed Downing, 'begin to embellish a country we know that order and culture are established' (Downing 1851: 53).

In the vanguard of exurban development were writers and artists seeking pastoral retreats within easy reach of their publishers and their urban patrons (Schmitt 1969). These were followed by a growing class of the well-to-do: professionals, executives and entrepreneurs, as well as those seeking a peaceful place for retirement. What they all had in common was a desire for country living without sacrificing accessibility to their urban sources of income, entertainment and consumerism. With the expansion of the railway network in the second half of the nineteenth century the rural hinterlands of the cities of the north-eastern USA were opened up for exurban development on a wide scale. Areas that had previously been occupied mainly by summer homes now became accessible on a daily, year-round basis (Muller 1976). In Canada, too, summer homes around Toronto soon became commuter settlements with the arrival of regular railway service in the 1870s (Punter 1974).

By the 1880s the commuter lines radiating from New York carried growing numbers of urban escapees out into the rolling woodland of Spectorsky's exurban wedge. High commuter fares discouraged most wage-earners and therefore helped to foster the area's image as an enclave for the well-to-do (Jackson 1985). The railway came to Bedford Village in northern Westchester County in 1847. In the next forty years it became a fashionable place for modest country estates, whose picturesque gentility to this day sets them apart from the suburban-like areas of a more recent migration (Duncan 1973). The completion of the Long Island Railroad, originally conceived as a direct route between New York and Boston, but with commuter service from the North Shore by 1860, led to the development of several exurban communities in what had previously been an area of large estates (Spectorsky 1955).

American railway companies and real estate developers collaborated to promote both the accessibility and the ambience of out-of-town locations (Jackson 1985). As early as 1849 there were 59 commuter trains arriving in Boston each day from 25 kilometres away or less and another 45 from further afield (Spectorsky 1955). Commuters' clubs became a regular feature of the trip to and from town, hence the origin of the 'club car' as the euphemism for first-class on North American railways. Indeed, the prestigious associations of exurban development prompted the development of a number of exclusive picturesque communities. Among the earliest of these was Llewellyn Park, a short railway trip from Manhattan in the foothills of New Jersey's Orange Mountains. It was developed by New York entrepreneur Llewellyn S. Haskell as 'a retreat for a man to exercise his own rights and privileges ... with special reference to the wants of citizens doing business in the city of New York and yet wishing accessible, retired and healthful homes in the country' (in Jackson 1985). Its distinctive features were a curvilinear road pattern (an explicit break with urban grid patterns) and a natural open space in the centre. Lot sizes averaged a little over three acres and a fifty acre 'Ramble' through the surrounding woodland completed the country atmosphere. Other developments along these lines were built over the next few decades. Because most of these were part of the garden suburb movement, they are examined in more detail in Chapter 5. Yet many of them were clearly an important element in the exurbanisation process.

In Britain the railway played an equally critical role in stimulating Victorian interest in exurban living. By mid-century the new lines radiating out of London and other large cities helped to consolidate the rising fashion among artists, writers and professionals for residing in the country. Such was the extent of this trend, that the landscape gardener Gertrude Jekyll, who moved to a cottage in the favoured region of the Kent–Surrey border, was prompted to complain of the steady influx of others to disturb her accessible rural retreat! (Marsh 1984). And accessible these areas certainly were. In the railway mania of the 1880s and 1890s scores of lines leading into urban termini were constructed. As in the USA, the commuter most valued by the railway companies was the first-class traveller living in the countryside and working in the city (Dennis 1984). Some of these wealthy exurbanites travelled some distance to their urban livelihoods. A surprising example of this is the report of businessmen from Manchester and Liverpool commuting from 'palatial villas' on the shores of Lake Windermere in the 1880s (Kellet 1969).

As suburbia proper, with the help of the electric tram, the underground railway and the omnibus, expanded, so too did the interest in residing in more rural surroundings. Cottages in the picturesque and arts and crafts styles were especially in vogue. Several new cottage colonies, such as Sapperton in Gloucestershire and Portmeirion in North Wales, established by writers and artists helped to popularise the cottage fad (Betjeman 1982). For those who preferred something more authentic, the continued exodus of farm labour from the countryside ensured a plentiful supply of cottages for renovation. While cottages may have attracted those of artistic and nostalgic bent, more conventional citizens gravitated to the exclusive villa estates which were springing up around villages on the outskirts of major cities. Around London the exclusive addresses of today's outer suburban ring – Chislehurst, Esher, Ascot, to name a few – were first established, and the 'Home Counties' began to take on a predominantly exurban character which, as the maps in Plate 3.4 show, spread rapidly across the region. Fine Edwardian villas transformed the appearance as well as the social life of villages. As in the USA, however, much of this residential development consisted simply of suburbia removed to the country. Builders promoted the benefits of combining suburban convenience with rural amenities. After the opening of the Metropolitan Railway station in 1892 the population of Amersham in Buckinghamshire (60 kilometres from central London) grew rapidly as estates of comfortable villas were developed (Coppock and Prince 1964). As a dormitory settlement, Amersham expanded steadily over the next fifty or so years. Its development at the turn of the century was repeated in numerous villages around London, and places like Gerrards Cross, Radlett, Haywards Heath, and Harpenden became familiar names on the destination boards of the London termini.

With the internal combustion engine the opportunities for exurban living reached their full potential. Previously restricted to locations within a carriage ride of a railway station, aspiring country dwellers now had more freedom in their selection of their ideal retreat. The motorised omnibus provided more efficient links to commuter rail lines, and by the 1920s commuting into town by car, a lifestyle that has come to define the character of the modern metropolis, had begun. In the USA, in particular, the automobile opened up the countryside for residential development on an unprecedented scale. New highways, many of them, like the parkways extending from New York City, dedicated exclusively to commuter and pleasure traffic, expanded the exurban radius (Newton 1971). The car also allowed for greater flexibility in the use of commuter rail services. In Britain the automobile remained a relative luxury during the twenties and thirties, and was used more for pleasure than as a means of exurban commuting. Of greater significance on a large scale was the spread of motorised bus services which encouraged new residential development in the countryside around most provincial towns and cities. Housing estates

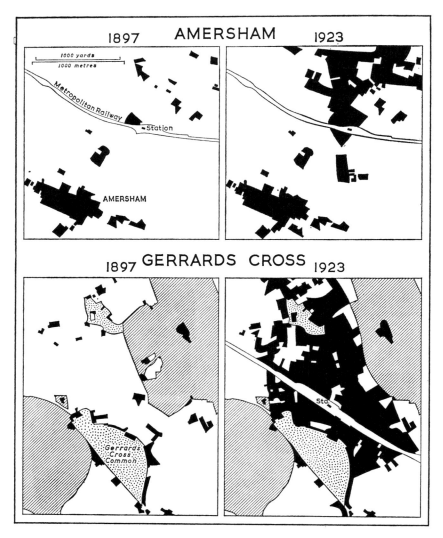

Plate 3.4 Rapid turn-of-the-century growth in the dormitory settlements of the Home Counties, from *Greater London*, ed. by J. T. Coppock and Hugh C. Prince, Faber and Faber, 1964

sprang up almost overnight around villages and along major roads leaving a legacy of inter-war ribbon development which prompted the first planning legislation to control haphazard building in the countryside, although not before many villages and large stretches of countryside had been transformed into bungalow-land.

After the Second World War, and accelerating since then, the steady trickle of urbanites moving into the countryside became a rapidly growing stream. Much of the evidence for this is to be found in what the current jargon of migration analysis variously calls counter-urbanisation, rural repopulation or, in reference to the reversal of historical patterns of rural–urban migration, population turnaround (Gilg 1986). It is a trend which has been recognised in most industrial countries. Perhaps the most remarkable reversal has been in the USA where, since the early 1960s, the population of non-metropolitan areas has increased more rapidly than that of metropolitan districts. As one would expect most of this growth has occurred in the metropolitan fringe, but rural areas beyond the fringe have also experienced disproportionately rapid increases in growth rates (Brown and Wardwell 1984). Small towns and villages in rural America, in particular, have experienced startling expansion rates since the early seventies (Johansen and Fuguitt 1984). Much of the repopulation of rural Britain has to do with the general drift of population from north to south (Gilg 1986). In a broad band covering lowland England and the Welsh Marches, rural population growth rates ranged between 5 and 15 per cent in both the periods 1961–70 and 1971–80 with the highest increases in the 'sunbelt' stretching from Norfolk through Oxford to Bournemouth. Even more remote rural districts such as mid-Wales have grown fairly rapidly (Champion et al. 1989; Gilg 1986).

The motives behind these trends are complex, for they are linked inextricably to the very process and character of modern urbanisation. The push out into exurbia and, more recently into rural areas beyond, is to some extent a reflection of the physical and demographic expansion of the metropolis. In this sense, migration from both city and suburbs can be seen as part of an overspill process, with the demand for affordable home-ownership as the driving force. Together with the decentralisation of economic and institutional activities, and the improvement of regional transportation infrastructures, this has produced new forms of metropolitan regions in which the distinction between rural and urban has become submerged by waves of urban-derived expansion. Over thirty years ago Gottmann coined the term 'megalopolis' to describe the sprawling band of urban, suburban and exurban settlement stretching from Boston through New York City and Philadelphia to Baltimore (Gottmann 1963). Similar regions on a smaller scale can be recognised in southern Ontario, southern California, and the Chicago area. Indeed decentralised sprawl characterises most North American metropolitan areas. Such has been the impact of

Plate 3.5   The scattered residential settlement of modern exurbia: just north of San Francisco

this that it has prompted an eminent interpreter of the American landscape to speak of a 'galactic metropolis', a metropolis which is everywhere (Lewis 1982). In many ways much of western Europe could be described in similar terms, perhaps with more reason. Certainly most of the conurbation of midland and southern England contains many of the essential character- istics of Gottmann's megalopolis.

In such regions the choice of where to live can be as much driven by the decentralising process itself as by a conscious desire to move to the country. Yet beneath these more general trends, the country life ideal still sets much of the tone of exurbia. It is at its most explicit in the continued enthusiasm for replicating the lifestyle of the country gentry. In the wide- spread absence of significant planning restrictions on this type of develop- ment there has been a sustained market for parcels of rural land which can be converted to small country estates (Healy and Short 1983). Scattered residential development of this kind has come to dominate rural landscapes around most cities (Bryant et al. 1982). Space, privacy, image and status are the hallmarks of the modern country estate, just as they were of their grander predecessors. Although eclecticism abounds, architectural styles and landscaping conventions are heavy in their emphasis on the creation of a gentrified or rustic enclave, shielded from the messier realities of rural life. Scaled-down versions of the English or Colonial country house and scaled-up versions of the pioneer log-cabin or traditional farmhouse, often available in pre-fabricated form, tend to dominate house design. And as we can see in Plate 3.6, the true romantic can even build an English cottage complete with authentic and fully fire-proofed thatched roof manufactered by Warwickshire Cottage Enterprises of southern California from reed imported from England! Landscaping generally follows the traditions of the picturesque, with plenty of ornamental plantings and wide expanses of manicured and chemically-perfected lawn, or the white-railed exclusiv- ity of equine pursuits. A sizeable land parcel located in natural surround- ings achieves both the sense of retreat and the status of property ownership that is the measure of exclusive country living.

Significantly tighter restrictions on residential development in open countryside in the post-war period in Britain have prevented the country estate sprawl which has intruded on so much of the North American landscape. Notwithstanding the limited supply, there appears to have been a healthy market for this type of property in recent years. A 1987 issue of Property Times, for example, reported a 'phenomenal demand' in the Home Counties for houses 'in the range of £200,000 and above, with pony paddocks, tennis courts and swimming pools', a demand which was becom- ing increasingly difficult to meet because of the trend towards subdividing larger estates into smaller parcels for more modest properties (The Times, 31 June 1987). That this demand has to be satisfied from within the existing stock of country properties, of course, is not entirely inconsistent

Plate 3.6   English thatched cottage, California-style!

with a countryside ideal which places great store in picturesque settings and in the cultural landscapes of the old squirearchy. Typical of the variety of much sought-after properties would be the following sample of estate agents' listings from a typical issue of *Country Life* magazine:

'Detached family home set in rural surroundings yet close to main line station . . . extensive gardens, grounds and paddocks about 5 acres.'

'Elizabethan manor House in a peaceful rural setting . . . in about 12 acres . . .'

'Superbly modernised period cottage of great character and charm . . .'
'Beautifully converted oasthouse in superb rural setting near old
world village . . .' [oasthouses were used for drying hops until
replaced by mechanical dryers – the aroma must be intoxicating!]

(Country Life, 2 July 1981)

From the family home, which like many properties of its kind would date
from the flurry of country retreat building that occurred between the
1880s and 1930s, to the converted oasthouse, the general tenor of these
advertisements and indeed of those which would be found in any recent
issue of *Country Life*, is one which emphasises the atmosphere of old rural
England. The market for traditional rural buildings, renovated or converted
to modern domestic standards, is particular strong. A trend which, as we
have seen, began at the turn of the century with the first wave of rural
depopulation, it has become not only a way of satisfying demand for
exurban property in a tight supply situation, but also demand for property
which matches one of the dominant images of the English countryside
ideal. Nothing conforms more to this ideal than a picturesque thatched
cottage. Such is the market for the perfect cottage that even the most
derelict command top prices. Renovated cottages, however, have been
joined by a wide range of other rural buildings which have been absorbed
into the exurban property market. To oasthouses we can add farmhouses,
barns, stables, mills, as well as more imaginative conversions such as abandoned schools, chapels, railway stations, and even lighthouses.

To a somewhat lesser extent this pattern has been repeated in North
America. As escalating real estate values and tighter planning controls have
reduced the accessibility of more conventional exurban estates and the
potential for acquiring parcels for new construction, the attraction of
the conversion and restoration of traditional rural structures has grown.
Farm enlargement has produced a steady supply of farmhouses for the
exurban market, while barns, school houses and churches appear to be
the main focus of residential conversions. Those seeking these kinds of
property need only to peruse the country living section of the *Sunday
New York Times* to find their rustic dream.

For those who lack either the resources or the desire for the exclusivity
of a small country estate or the rustic authenticity of a converted barn,
developers and builders have been only too willing to provide a more
standardised and ready-made alternative. Speculative building of upper-
income housing developments has been a feature of exurbia, as we have
seen, since the middle of the nineteenth century. The modern version is
generally a continuation of this genre. In North America subdivisions of
large homes on substantial lots (usually up to a couple of acres) dot the
countryside around most cities. In his study of the Kentucky Bluegrass
Plain around Lexington, Patel identified over one hundred such subdivisions

(Patel 1980). These 'deep suburban developments', as one observer has characterised them (Dorst 1990), are heavy on contrived rural imagery, not only in their architecture and landscaping, but also in their promotion to potential buyers. Subdivision names like 'Willow Lane', 'Amberview Village', 'Whisper Woods', 'Cedar Mills', and 'Walden Spinney', create caricatures of idyllic natural settings, harmonious rural communities, rustic charm, leisurely lifestyles and the gentrified exclusivity of country estates.

Villages and small towns on both sides of the Atlantic are strong magnets for exurbanites seeking their vicarious experience of country living. Not only do they provide ready access to essential services, but they also conform to notions of the ideal place to live. The demographic revival of small towns across North America has been fuelled by the influx of people looking for a gracious old mansion, a vacant lot on which to build their dream home, or a peaceful subdivision on the edge of town. As Dahms has observed in the Canadian context, the small town and village have become the focus for a highly mobile and decentralised exurban population (Dahms 1988). In Britain, as we have seen in Chapter 2, the village has attained the status of a national cult. Newcomers to Akenfield, the Suffolk village made famous by Ronald Blyth's study, are surrounded by

> evidence of the good life, a tall old church on the hillside, a pub selling the local brew, a pretty stream, a football pitch, a handsome square vicarage with a cedar of Lebanon shading it, a school with jars of tadpoles in the window, three shops with doorbells, a Tudor mansion, half a dozen farms and a lot of quaint cottages . . .
>
> (Blyth 1969: 17)

The more picturesque and accessible the village, the more prestige it acquires as a desirable exurban address (Connell 1978).

The importance of traditional village character to adventitious exurbanites is reflected in their domination of village preservation and amenity movements. As Chapter 6 reveals in greater detail, the rapid growth of these movements on both sides of the Atlantic in recent years is largely attributable to the influx of middle-class urbanites whose aesthetic sensibilities and vested interests converge around strongly preservationist ideologies. The desire to keep the village as it is – to preserve its traditional appearance and to protect its surroundings from further development – is, of course, entirely consistent with the values of those who have invested in the village and, indeed the countryside in general, as a residential ideal.

This preservationist mentality is a measure of the extent to which the countryside ideal has penetrated the culture of exurbia. The principal aim of preservation is the maintenance of landscapes and communities which have been fabricated by exurbanite lifestyles and values. This is at its most explicit in the NIMBY-ism ('Not In My Backyard') which protects private amenity – space, seclusion, pleasant vistas, arcadian settings – and residential exclusivity from undesirable land uses and people. Yet it is also consistent with the emphasis on the outward display of the style of country living; in the attempts to replicate the landscape of the country gentry or the quaint atmosphere of old country life. The latter has become a particularly popular aspect of exurbanite living in recent years. In a reinvocation of arts and crafts aesthetics, the search for rural authenticity in the replication of country vernacular extends from the perfect cottage restoration to interior design, gardening and cooking. The tone of this is captured in the plethora of magazines devoted to the promotion of country lifestyles, which were discussed in the previous chapter.

While these magazines are aimed at a wider readership than exurbanites, in their direct reference to the country living experience they do reveal the extent to which exurbanites have established their own version of rural life. Of course this does not extend to all exurbanites, many of whom have more utilitarian demands on the countryside. Yet it does reflect the values of that element of exurbanite society which has given self-conscious and tangible expression to the countryside ideal in its private landscapes. That this has spilled over into the public landscape is evident in the re-fashioning of exurban places around a culture of country-style consumerism. In rustic inns and country restaurants, antique shops and craft boutiques, village fairs and summer music festivals the culture of exurbia has become a profoundly consumer experience. It is woven seamlessly into a preservationist ethic in which the creation of rural authenticity goes hand in hand with commercial opportunity. In a recent study of an exurban community in Pennsylvania, Dorst argues that this is representative of the culture of advanced consumer capitalism; of the culture of post-modernity reflected in the transformation of the self-inscribed nostalgia of exurbia into a commodity form (Dorst 1990).

BACK TO THE LAND

In the discussion so far I have concentrated exclusively on the place in the country as a partial escape; as a means of having the best of both urban and rural worlds. Running through the history of the country place there is a strong current of romantic idealism about nature, land and country life. For the most part this has not been translated into a detachment from urban society or into a full commitment to a rural lifestyle. Yet, as we have already seen, throughout the rise of modern urbanisation there has been a persistent undercurrent of ideological rejection of city in favour of country life. This has received tangible expression in a variety of attempts to seek permanent rural alternatives to the urban–industrial system; in other words to make a complete break with urban life and to go 'back to the land'.

Insofar as we can speak of a back-to-the-land *movement* (Marsh 1984),

Plate 3.7 The gate of success: late twentieth-century gentrification of Toronto's countryside

it is one which has been dominated by the history of alternative communities. The latest version of this form of escape is the modern rural commune which first took hold of the youthful imagination in the late 1960s. But the sixties commune has antecedents which stretch back over at least one hundred and fifty years of attempts to establish radical, land-based alternatives to conventional urban living. The idea of natural rights to land as a political reaction to agrarian and industrial capitalism was particularly strong in the late eighteenth and early nineteenth centuries. Indeed, as early as 1649, the Digger movement under Gerald Winstanley had attempted what was arguably the first co-operative land settlement at George Hill in Surrey (Hardy 1979). A century and a half later, with the Land Scheme ideas of Paine and Spence and then the rise of Chartism and Owenism in the 1820s, the back-to-the-land movement began to take hold on both sides of the Atlantic. Although the Chartist movement collapsed as political force, the Land Company eventually attracted 70,000 subscriptions from all over the country (Hardy 1979).

In the USA the establishment of New Harmony by Robert Owen in 1825 marked the beginning of a long period of experiments with utopian communities. Many of these were sectarian in origin. For them a land-based settlement was the only possible context for their rejections of conventional society. But several communities, rather than being driven primarily by religious ideology, explicitly embraced the undercurrent of antipathy towards the expanding system of urban–industrial capitalism. At the 'New Harmony Community of Equality', village and farm life was seen as the proper setting for the communistic ideals of Owenism (Fogarty 1972). New Harmony failed within two years but it remains an influential event in the history of American utopianism and in the ideology of the back-to-the-land movement. Another and perhaps even more significant experiment was the Brook Farm Community. Set up in 1841 by George Ripley 'to insure a more natural union between intellectual and manual labour' (Melville 1972), an ideal which Ripley believed could be achieved only by making agriculture the basis of community life, Brook Farm became the focus of the intellectual anti-urbanism and transcendentalism of the period.

While the search for agricultural utopia, under the dominant influence of religious sectarianism proceeded westwards with the spread of settlement in the USA, the second half of the nineteenth century saw the rise of a strong back-to-the-land movement in Britain. Jan Marsh's book (1984) provides a fascinating and detailed account of this movement, and I make no apologies for drawing on it for the basic structure of the brief summary presented here. Driven, as in the USA, primarily by the rising anti-industrial ideology of the times, it was a movement which resulted in a variety of attempts to re-settle rural Britain. In the early stages it was dominated by communal experiments; a continuation, after a fashion, of the radical land-based ventures earlier in the century. Among the first of these experiments was that of St George's Farm at Totley, just outside Sheffield. Established by Ruskin to put into practice the principles of his Guild of St George, the Totley community set out to restore lost values through a return to an economy based upon agriculture and crafts. As time went on, Ruskin became more interested in laying out botanical gardens on the site and the community itself, under the influence of Edward Carpenter and the leadership of the Sheffield socialist John Furniss, shifted to more co-operative principles (Armytage 1961; Marsh 1984).

The next thirty years saw a proliferation of experiments to put people back on the land. A large proportion of these were organised along communal or co-operative lines – 'land to the people' communities. An early example was Graig Farm at Millthorpe in Surrey, which, under the influence of the ideas of Thoreau and Edward Carpenter, was organised to provide a combination of manual work and intellectual creation (Marsh 1984). Another community was the Whiteway colony in the Cotswolds which was established in 1883 by the Fellowship of the New Life which subscribed to the idea of a free union of cultivators. Whiteway attracted many who held socialist and libertarian views including Havelock Ellis, Olive Schreiner, Ramsay MacDonald and the apparently ubiquitous Edward Carpenter. It survived until the 1930s when it became a settlement of independent cottagers. By the end of the century the more radical ideas of Kropotkin were being put into practice in communities like the Clousden Hill Communist and Co-operative Colony on the outskirts of Newcastle in 1895 (Marsh 1984).

These experiments in agrarian socialism, however, were less popular than the looser co-operatives and cottage farming ventures of the time. One of these, the Methwold Fruit Farm Colony, was begun in Norfolk between 1889 and 1890 by R. K. Goodrich whose plan was to colonise a parcel of land with small-holders and to sell produce direct to the consumer. His advertisements in the national newpapers brought enough applications to settle fifty families on 2- to 3-acre plots within ten years. The virtues of the Methwold Colony were amply praised by the *Cable* in its editorial:

> ... Methwold ... a new order of things has been inaugurated. The land there is being taken possession of not by the country folk, but by clerks and tradesmen from London and other large centres of population ... It points to the fact that agriculture is man's natural occupation and that, in many cases, the love of it is inherent
>
> (in Marsh 1984: 115)

Several other schemes in similar vein to that at Methwold were initiated at the turn of the century. One of the most widely publicised of these was Mayland. It was begun by one Thomas Smith, a Mancunian who was inspired by Robert Blatchford's series of articles in 1895 in *The Clarion* calling for people to return to the land (Hardy 1979). Smith acquired almost 5 hectares near Althorne in Essex and advertised for fellow settlers, offering individual ownership with voluntary co-operation. The idea was to sell produce to the London market. Under the patronage of Joseph Fels, an American soap manufacturer who had come to reject the industrial system and took to dispensing his money to schemes which offered an agricultural alternative to urban occupations, Mayland grew to 250 hectares by 1905 with 21 holdings as well as a village.

The land settlement idea reached its zenith in Britain during the Edwardian period. Several books were published on the subject, including A. C. Fitfield's 'The Cottage Farm Series' between 1906 and 1909, and a series by F. A. Morton entitled 'Life on Four Acres'. Perhaps the most popular author of all in an era saturated with rural nostalgia was F. E. Green. For Green the virtues of settling and surviving on a small-holding in the country were self-evident. 'When we have removed from our eyes the shine and grit of our cinder-strewn cities,' he wrote in *The Awakening of England*, 'then perhaps we shall awake to the necessity of re-colonising our own land' (Green 1912: 369). While planned colonies like those at Methwold and Mayland were small in number and relatively short-lived, Green's ideas on independent small-holdings have had a lasting influence on the British countryside. The Small Holdings Act 1907, which improved the position of tenants in relation to landlords, increased the opportunities for pursuing small-scale agriculture. There were even co-operatives of small-holders such as the Reigate Small Holders Ltd, which leased 194 acres at Leigh from Surrey County Council (Green 1912), but, in the main, small-holdings were individual ventures of what Marsh calls 'cottage farmers' (Marsh 1984).

The return to the land, however, was seen also as a direct way of alleviating urban social problems, particularly unemployment and vagrancy. In 1887 the Mansion House Inquiry into the Condition of the Unemployed recommended the idea of settling them on under-used farmland. In 1892 the Reverend Herbert Mills started The Home Colonisation Society which established farm colonies for the urban unemployed. At its first colony at Starnthwaite in the Lake District, however, the communal, arts and crafts character of the colony attracted intellectuals and radicals in addition to the unemployed (Marsh 1984). Of sterner stuff were the farm colonies established by the Salvation Army. In his campaign for the 'elimination of the submerged tenth', William Booth, the Army's founder, proposed 'removing men from the city and setting them to work on the land' (in Marsh 1984: 127). At the first site of the Scheme of National Land Settlement at Hadleigh in Essex, unemployed men were subject to a strict regime and taught the basics of agriculture in preparation for emigration to the colonies. Rider Haggard (1905) described it as 'More like a Poor House colony for destitute men' (126). In fact, rural resettlement as a solution to urban destitution was commonly used during the early years of the century by Poor Law Authorities and philanthropic societies.

The Salvation Army was active also at this time in the USA. In his report to the British government, Rider Haggard compared the Army's colony at Fort Amity, Colorado most favourably with the conditions at Hadleigh. At Fort Amity, he wrote that there was 'a population of about 275 persons living in happiness, health and comfort, and, in sundry instances, comparatively wealthy' (Haggard 1905: 69). Indeed it was more a homesteading scheme than a poor relief colony. Back-to-the-land homesteading was the American version of the British small-holding settlement. Again, in part at least, it was a product of turn-of-the-century concern about urban living conditions. In 1914 an article in the U.S. Department of Agriculture Yearbook entitled 'The Movement from City and Town to Farms', indicated official support for a back-to-the-land movement. Other articles, such as Lyman Beecher's 'Training City Boys for Country Life' promoted the idea of settling unemployed youth on the land (Bowers 1974). Beyond this, however, there was a campaign for more general relocation of urban residents in the country, associated in part with the broader processes of exurbanisation. The benefits as well as the feasibility of self-sufficiency on small-holdings were widely promoted in the 1920s and 1930s. A particularly influential figure in this movement was Ralph Borsodi who in his book *Flight From The City* (subtitled 'The Story of a New Way to Family Security') described his own family's experiment with back-to-the-land homesteading. Borsodi provided practical advice on small livestock rearing, vegetable gardening, the processing and storage of produce as well as the economics of self-sufficiency (Borsodi 1933).

It was no coincidence, of course, that the back-to-the-land movement gathered momentum during the depression years. Thousands of the urban unemployed resorted to subsistence farming as a means of survival in the early 1930s. The Department of the Interior established a Division of Subsistence Homesteads in 1933, with a mandate to establish projects across the country (Shi 1985). The Dayton Homesteading Scheme for example, consisted of 3-acre plots on a site just 3 miles outside Dayton, Ohio and was settled by thirty-five families drawn from the unemployed. Each family was to build its own house and 'become as nearly self-sufficient as were the pioneers of 100 years ago' (Borsodi 1933). The scheme, however, did not involve a complete break with urban civilisation for it was located sufficiently close to the city to enable the homesteader to commute to work, schools and entertainment. Like most homesteading schemes it failed in its original objectives because few urbanites took to the idea of subsistent living.

The Dayton Scheme and other agrarian solutions to the Depression came at the end of an approximately fifty year period of experimentation with back-to-the-land communities. Some of these, like the Salvation Army colonies, used the land as a practical panacea for urban social problems. For others, like the sectarian communities which I have not discussed here, an agricultural way of life was often the foundation of their religious ideologies. Yet at the heart of the back-to-the-land movement were communities which were motivated by 'an holistic view of an alternative society' (Hardy 1979: 81), in which the escape to a rural yet communal existence was seen as the only way to ensure the fulfilment of human potential and even the survival of civilisation. And it is this same philosophy which has provided the impetus for the alternative community movement of the last forty years or so. This was epitomised by the experiments with pacifist communes in Britain in the 1930s and 1940s. The driving force behind these was John Middleton Murray and his Christian socialist belief in the reconciliation of capitalism and socialism in a society founded upon farming communities. By the end of the 1930s there was a small network of pacifist communes in the British countryside (Hardy 1979).

There is a direct philosophical link between these communes and those of the sixties and seventies. The so-called 'hippie' communes which first appeared in the late sixties were, in essence, the symbols of the counterculture, of a radical pacifism and a dissaffection with the whole industrial/scientific/capitalist/consumer structure (Roszak 1969). As we saw in Chapter 2, this ideology was intertwined with the new environmentalism and the search for an alternative and sustainable lifestyle. As a purely political statement communes were probably more effective in the city than in the country. However, although urban communes were established, it was the urge to 'drop out' which became the driving force of the movement. And so, armed with the writings of Thoreau and Leopold, the Whole Earth Catalogue and the quotations of Timothy Leary, the people of the movement left their cities (or, more commonly, their middle-class suburbs) in search of rural land. Although some rural communes were established in Britain, the movement took its lead from North America. Communes were founded in every region of the USA except (not surprisingly!) for the South and the Middle West (Melville 1972). They spread also to Canada, especially to British Columbia and Ontario which not only had a plentiful supply of cheap backwoods land but also provided a haven for American draft-dodgers.

The heartlands of the rural commune were, however, California (where it all began), New Mexico and New England. At one level there were the 'hip' communities, which were in the mainstream of the drop-out culture. Founded on a potent mixture of mysticism, anarchy and socialism, these were the experiments in self-actualisation and individual freedom, which were scattered across the Californian and New Mexican countryside by the early seventies. At another level, as re-invocations of nineteenth-century transcendentalism, communes were holistic experiments in a self-sufficient lifestyle in harmony with nature (Berger 1981). Melville illustrates this with the example of Ray Mungo (Melville 1972). Mungo was a radical student leader at Boston University who dropped out in 1968, set up a commune in Vermont, and became one of the leading gurus of the commune movement:

The word Vermont popped into heads almost simultaneously. Vermont! Don't you see, a farm in Vermont! A free agrarian communal nineteenth century wide-open healthy clean farm in green lofty mountains! A place to get together again, free of the poisonous vibrations of Washington and the useless gadgetry of urban stinking boogerin' America.

(in Melville 1972: 81)

Here was the rural commune as 'the contemporary bearer of the pastoral tradition' (Berger 1981). And in this sense we can recognise a third element in the modern commune: the commitment to alternative technology, including organic farming, small-scale industry, crafts, and renewable energy sources. Most rural communes espoused this 'whole-earth' philosophy, but probably the most durable have been those for which it was the basis of a viable group economy. (In contrast to the more ephemeral experience of the more socially and spiritually-based communes.) The survival, and indeed the continued formation of practical back-to-the-land experiments is to some extent a function of their diversity. They vary from those which have been in the mainstream of the counter-culture to quasi-scientific experiments in environmental sustainability.

While communal and co-operative experiments have defined much of the ideology of the back-to-the-land movement, quantitatively they have had less impact on the countryside than have individual back-to-the-landers. As we have seen, even at the height of the community-based movement earlier in the century there was an equally strong crusade for independent family small-holding and homesteading. Much of this earlier attraction to small-holdings came from those who could still keep one foot in the city. For example, alongside (and thoroughly condemned by) Borsodi's ideas of subsistent small-holding, ran the commuter agrarianism of the country life movement, led by Liberty Hyde Bailey (Shi 1985). And much of the small-holding movement in Edwardian Britain involved those who maintained urban employment. Dabbling in farming as a sideline or source of relaxation has since become a common phenomenon in exurban areas. Hobby-farming, as it somewhat derisively termed, is a diverse activity, ranging from large gentleman farms to a few exurban acres. For many who indulge in hobby-farming, either on a small or a large scale, the principal objective is often to obtain the tax benefits which come from establishing some kind of agricultural activity. This reduces the cost of maintaining an exurban lifestyle while at the same time permitting the creation of the illusion of going back to the land. In this respect hobby-farming is an integral element of the landscape of exurban estates (Punter 1974).

The extent to which hobby-farming represents a desire to go back to the land is therefore open to question. This is not to say that it does not embrace a genuine desire to farm. Certainly the maintenance of small livestock operations which is common amongst hobby-farmers demands genuine commitment. Moreover, the growing interest in flavouring exurban living with the taste of rural authenticity extends to the revival of small-scale and traditional ways of farming. Much of this is associated with the increase in demand by exurbanite society for produce fresh off the farm. Free-range eggs, organically-grown vegetables, hand-made butter and cheese have become an essential part of the exurban lifestyle. Finally, in some parts of North America at least, acquiring a few acres of urban fringe land on which to re-live their rural roots and produce traditional food has become a common practice amongst immigrants from southern Europe and South-East Asia.

It is often difficult to draw a distinction between exurbanites who dabble in agriculture as a hobby and those who go back to the land completely. Amongst the former there have always been individuals who become deeply involved in their farming and in the life of the local farming community. Amongst the latter a complete break with urban life is rare, especially with modern telecommunications. For the most part, however, the true back-to-the-lander has sought remoter pastures than those of exurbia. There is, of course, a romantic mystique surrounding the idea of abandoning city life for the pleasures of a farm. This has been fostered, as we saw in the previous chapter, by the many personal accounts of the farm retreat experience which have been published over the past century or so. In the absence of any systematic study of the phenomenon, these accounts provide with us the best available insight into both the ideal and the reality of going back to the land. They contain, as Schmitt has said of their early versions, a mixture of romanticism and practical wisdom (Schmitt 1969). Among the many examples of this genre that could be cited, perhaps the two that most vividly capture the contemporary back-to-the-land experience are those by Noel Perrin and Jeanine McMullen. In his four books Perrin (1978, 1980, 1984, 1991) describes his abandonment of city life for that of a Vermont farm. With each book his initial romantic-isation of living the simple life becomes increasingly tempered with the experience of scratching out a living from the land, and culminates in his ambivalence about it all. McMullen's story, *My Small Country Living* (1984), is typical of the many contemporary British accounts of urbanite experiments with farming. Long on ironic humour about the daily realities of hill-farming in Wales but also on the delights of the Welsh countryside, McMullen summarises her farming experience as 'a paradise, a hell on earth, an absorbing passion and a thundering great millstone around my neck.'

What both these books illustrate is the gap between ideal and reality which has always dominated back-to-the-land philosophy. Utopian notions of the simple life, collective or individual, have generally come face to face with the difficulties of surviving on the land, especially in the marginal farming areas, like Vermont and the Welsh Hills, to which back-to-the-landers have tended to gravitate. Perhaps the real reason why there have been so many books written about the experience is that it was the only way for the new pioneers to make a living! Perhaps, too, the only successful back-to-the-landers have been those with means of support other than farming. Writers, artists, potters, weavers – the arts and crafts culture in general – have long dominated the movement. The farm is thus a romantic locus for art and whatever income it can bring.

In the process of realising their own particular version of a country retreat, the country gentry, exurbanites, weekend cottagers, even back-to-the-landers have profoundly altered the character and the meaning of the rural landscape. They have fabricated a landscape which has transformed both natural environments and productive spaces into areas which conform to the idealisation of countryside as a place of leisure, refuge and alternative living. For the most part it is an amenity landscape, designed to provide pleasure rather than economic sustenance. It is also a predominantly private landscape controlled by the power and exclusivity of property ownership. It is therefore one in which the attraction of the countryside has taken on the form of a commodity, with land acquiring market value based upon its amenity rather than its productive value, a shift which has spilled over into the general rural economy. The essence of this analysis is that the country place phenomenon creates its own culture which it imposes on landscape and community alike. In some instances it has become the dominant cultural force, displacing traditional rural activities with a process of gentrification that has converted whole communities into amenity-based residential settlements. Yet it also offers the prospect for the revitalisation of rural communities through their integration into the economic and social fabric of exurban landscapes.

4

THE PEOPLE'S PLAYGROUND

Alongside the creation of the private amenity landscapes of country retreats, ordinary folk have shown an increasing interest in using the countryside for leisure, to the point that it has become the playground of society at large. Through the simple enjoyment of its scenery and open space to its use for a variety of outdoor recreational activities, the countryside has acquired a popular and tangible significance which has done much to sustain its idealisation in the public mind. Such is the strength of this perception of the countryside that it has generated increasing pressure on rural land and natural environments, as well as a growing demand for the protection of its amenity value and its public accessibility. This has resulted in the designation of rural and wilderness areas as leisure environments. It has also been accompanied by the rise of a substantial countryside recreation industry which has further stimulated public interest in its amenity use.

Until about the middle of the nineteeth century, the use of the countryside for relaxation and pleasure was the preserve of the affluent. For the most part recreational pursuits were followed within the broad confines of country estates. However, during the eighteenth century there was a growing fashion for touring the English countryside. As Ian Ousby's recent book has shown in fascinating detail, encouraged by improvements in transportation, and aided by an emerging industry in maps and guidebooks describing places of interest, the rising middle class took to the road in growing numbers (Ousby 1990). Country houses were a major attraction. As they became more spectacular and ostentatious in their architecture and their landscaping, many, like Chatsworth, Blenheim, Castle Howard and Wilton, acquired the status of show houses, regular tourist sites not only to be viewed from a distance but also to be entered and appreciated from within. The owners of the more popular country seats experienced the same problems with tourists as their descendants do today. By 1760 the Duke of Devonshire was designating 'two public days in the week', and by the 1780s Blenheim was receiving visitors only between 2 and 4 p.m. By the early nineteenth century some owners were beginning to charge admission!

Country houses, however, were not the only item on the tourist itinerary. Indeed, with the cult of the picturesque and then the rise of romantic attitudes to nature, wild and rugged scenery became the principal object of the traveller's attention (Ousby 1990; Thomas 1983). For those with the affluence and the fortitude to embark on the Grand Tour, the Alps were an essential part of the trip. But increasingly it was wilder areas closer to home which drew most English tourists in the eighteenth and the early nineteenth centuries. Writers, artists and intellectuals were in the vanguard of this trend. The artistic conventions of the picturesque, proselytised most notably in the context of the actual landscapes of the Lake District by William Gilpin, sent hordes of amateur sketchers in search of the perfect scene. Descriptive guides like Richard Warner's *Tour through the Northern Counties of England, and the Borders of Scotland* (1802), and Wordsworth's *Guide through the District of the Lakes* (1822), led people not only to the major scenic spectacles, but also to a way of appreciating the landscape. Warner put the Peak District on the tourist map by describing the seven 'Wonders of the Peak' which included the Castleton caves, Malham Cove and Goredale Scar. And, of course it was the Wordsworths who did most to popularise the Lake District as an early tourist mecca. By the 1830s it was the most popular spot on the summer itinerary of the leisured classes (Ousby 1990).

The spread across the Atlantic of picturesque and romantic ways of appreciating the landscape prompted an equal enthusiasm for leisure travel. By the mid-eighteenth century the more adventurous of the middle classes could take carriage journeys from the cities of the eastern seaboard into the wilder areas of the interior and travel as far even as Niagara Falls (Huth 1957). By the 1840s people were able to journey by steamboat up the Hudson River to Albany and from there into the Catskills (which became something of a fad), and on into the Adirondacks and the White Mountains. Closer at hand, natural features such as Pasaic Falls for New Yorkers and the Natural Bridge for Baltimore citizens became sightseeing spots. There were books promoting travel, such as Willis's immensely popular *American Scenery*, published in 1840, Charles Lanman's *Summer Books*, and of course the influence of Burroughs and Thomas Cole. Indeed Cole was active in organising excursions of artists and writers into the wilderness, and by 1850 travelling through the countryside, viewing and painting scenery had become a popular pastime for the leisured classes (Schmitt 1969).

THE RISE OF MASS RECREATION

In the pre-railway years travelling in the countryside as a tourist remained very much a minority taste. And of course it barely touched the mass of the population. There is some evidence that by the 1830s the working class of British cities was using the surrounding countryside for relief from an alien industrial environment. The countryside was still usually only a short walk out of town, and so factory workers escaped there during their limited leisure time, 'not to look at nature with the dreaming gaze of poets, but to regain good fellowship amidst the mountains and dales away from the antagonistic relationships of the factory' (Hill 1980: 15). Most of these workers were recent migrants from rural areas, and they must have craved the openness of the countryside. In his evidence to the Public Works Committee in 1833, Dr Kay, the great urban health reformer, reported that 'thousands of men and women left the rapidly growing squalor of the industrial towns in an effort to find healthful exercise in the open air' (in Hill 1980: 15). A growing factor in this was the disappearance of urban open spaces as the new industrial cities grew. In Sheffield, for example, there was no common land left by 1845. The appeals for public parks had been ignored, so many were prepared to walk for several hours on a Sunday to reach the open moors (Pollard 1959).

In these early years, however, countryside recreation was still an ill-formed idea. Amongst the adventurous rich and the artistic and literary community it was very much bound up with romantic ideals, and for more ordinary folk it was constrained by their relative lack of mobility and the absence of any concept of regularised leisure for the working class. By mid-century, however, conditions began to favour the rise of more general recreational use of the countryside. Two factors were responsible for this: changing attitudes to leisure and improvements in transportation.

In Britain until about the 1850s leisure was largely the privilege of the affluent. There was a clear distinction between the so-called leisured classes – those whose income did not depend upon continuous work – and the rest of society. The Protestant work ethic, with its repressive combination of puritanism and capitalism ensured the acceptance and persistence of this distinction throughout the early years of the industrial revolution. In the eighteenth and early nineteenth century factory owners simply got as many hours a week out of their workers as they could. Conditions were not much better for the agricultural labourer. Even those who were higher on the social scale, such as tenant farmers, tradesmen, clerks and others on the fringes of middle-class status usually worked long hours, with only Sundays as a regular relief from work. Fundamentalist attitudes towards the sabbath ensured that even this one 'free' day of the week was not intended for idle or hedonistic relaxation. This did not mean that people did not find time for such pleasures. Indeed in many of the early industrial towns absenteeism was common, particularly on the day after the Sabbath, which became affectionately known as 'St Monday' (Walvin 1978).

Cunningham (1980) attributes the first signs of change in this situation to the end of what was a period of crisis in the industrial revolution, a period, that is, of growing social concern over public health and the working conditions of the factory system. This led to a shift in attitudes towards the relationship between work and leisure which culminated in legislation to regularise working hours. By the 1870s bank holidays and half-day holidays on Saturdays were becoming formalised. The budding trade union movement played a central role in persuading both government and employers of the importance of providing weekly leisure time for the working class. But Victorian philanthropy was also a significant factor in promoting the social value of regular relief from the working week as well as the virtues of achieving this by escaping the city. In 1892, for example, the Fresh Air Fund was launched by the philanthropist Sir C. Arthur Pearson to give deprived children outings to the seaside and the country-side. By 1909 the Fund had financed day trips for more than two million children (Walvin 1978).

The railway age

Regularised leisure periods gave people the opportunity to get out into the countryside. The railway provided them with the means of doing so within the short space of a Saturday afternoon or a Sunday holiday. Before the railway, only the countryside within walking distance of their homes was accessible to working people, and as cities expanded this countryside became increasingly inaccessible. The high cost of early railway travel prevented the immediate invasion of the countryside by hordes of city dwellers. However, by the 1860s cheaper fares and more improved service encouraged growing numbers to take day-trips out of town. The railway companies, too, were instrumental in promoting the idea of informal rec-reation in the countryside. Seaside resorts were, from the beginning, the most popular destination, but excursion trains travelled on summer week-ends from the textile towns of Lancashire into the Lake District and from Sheffield into the Derbyshire Peaks (Walvin 1978).

The railways, in fact, were an indispensable element of the earliest of popular countryside recreation activities, namely rambling. By mid-century, this had already become a pastime of the well-to-do, with clubs like the Sunday Tramps. However, within twenty years it had become a mass-pursuit, a socialist-inspired people's movement, no less, which campaigned for open access to the common moorland which had been appropriated for the sporting pleasures of the landed gentry (Hill 1980). This movement is discussed in more detail in Chapter 6. Suffice it to say here that, by the end of the nineteenth century country rambling was a highly organised

Plate 4.1 Railway Rambles poster, 1935

Plate 4.2   Discover the Countryside: British Rail/Countryside Commission
campaign poster, 1987

activity with a large following. Northern working-class rambling clubs not only campaigned for access but also organised excursions by rail into areas like the Welsh Mountains and the Pennines. Rambling clubs were formed in the south too. Several of these were associated with the access lobby. By 1905 a national Federation of Rambling Clubs had been formed. The railways were essential in providing access to areas attractive for rambling. Clubs negotiated special fares and organised excursions for members, and rambles were planned around train schedules and routes. One famous London rambler, E. S. Taylor compiled a series of rambles which linked all the railway stations in the Home Counties (Hill 1980). And, as Plates 4.1 and 4.2 show, the railways have continued to be involved in promoting the enjoyment of the countryside throughout this century.

In North America the railway played a similarly important role in the popularisation of countryside recreation during the nineteenth century. What is less clear, however, is the influence of changing attitudes to leisure. Historians have emphasised the dominance of the Puritan work ethic in Colonial America and its intensification with the rise of industrial society. Much of this simply reflects the imperatives of a pioneering and individualistic society: work bred success, idleness failure. This became a convenient justification for the appalling exploitation of new immigrants in the factories and sweatshops of American industry until well into this century. It was also the reason why legislation to control working hours and to guarantee leisure time was much slower to appear than in Britain. Yet as the nineteenth century progressed, and as society became increasingly urbanised, considerable emphasis was placed on the importance of leisure and recreation. The urban parks movement was instrumental in establishing the need for accessible recreation space for the urban working class, and social reformers were beginning to raise the public conscience about industrial working conditions.

Despite what appears to be a sluggish move towards an acceptance of leisure time for ordinary working people, countryside recreation was by the middle of the nineteenth century an increasingly popular form of relaxation. If attitudes to leisure for the working class were slow to change this was not so for the growing numbers who counted themselves amongst the urban middle class. It was this group which could shift the work ethic down the social ladder and establish patterns of work which guaranteed regular and extended periods of leisure. And so it was the middle class which began to use the countryside for recreation on a large scale. Much of this was fostered by the continued interest in nature and the outdoors. Outdoor sports grew in popularity and with them came the invention of the country club, which as we saw in the previous chapter soon became an integral part of the social life of exurbia. For the less active urbanites who wished only to enjoy the visual qualities of the landscape or who sought the therapeutic resources of the countryside, resorts such as Saratoga Springs opened up on a grand scale (Huth 1957).

By the 1860s, then, what Hans Huth (1957) has termed 'the summer migration' was in full swing from the cities of the eastern seaboard. As in Britain, railroad companies played an active role in promoting excursions into the countryside and into wilder country beyond. In 1857, for example, the Baltimore and Ohio Railroad advertised a special excursion train to take writers, painters and photographers on 'The Picturesque Line of America' (Huth 1957). But the railway was also the catalyst for more general leisure travel. Day-trips into the countryside became a popular form of recreation for urban residents, expanding existing resort areas and opening up new ones virtually overnight. However, the railway's main contribution was to provide access to the more remote and spectacular areas of the American landscape. This was a particularly significant development because it marks the beginning of what has become a dominant theme in outdoor recreation in North America as a whole: the cult of wilderness. As we have already seen, the notion of travelling in order to appreciate the sensory and spiritual pleasures of nature was already well established. But it was the growing numbers of ordinary people (as opposed to artists and intellectuals) who were able to put this notion into practice that led to its cultural entrenchment.

By the end of the nineteenth century many of today's patterns of countryside recreation had begun to emerge. Increased leisure-time, affluence and mobility played leading roles in this, but there can be little doubt that these were factors which served to satisfy and further stimulate an established demand for fresh air, open space, nature and country atmosphere. In many ways the countryside had already become a mental escape for urban society. With the growth in its recreational use it also became a physical escape, increasingly perceived as a public amenity. However, as long as people depended upon nineteenth-century modes of transportation there were clear spatial limits to their enjoyment of the countryside. The railway certainly revealed the potential for countryside recreation but it was only with the invention of the internal combustion engine that this potential could be fulfilled.

Leisure and the automobile

If the automobile, as we saw in the previous chapter, expanded the possibilities for weekend cottaging and exurban living then so much the more did it raise the potential for casual and unlimited countryside recreation. Cars began to appear on the roads around the turn of the century and for a decade or so their cost and unreliability limited their ownership to the wealthy and the adventurous. Nevertheless, by the early 1900s automobile touring was becoming a fashionable activity and the countryside was its focus. Indeed for the first twenty or so years of the century the motor car was designed and marketed primarily as a pleasure vehicle (Pettifer and Turner 1984). Early advertisements, as Plate 4.3 illustrates, stressed the joys of travelling the open road and frequently displayed new models against a pastoral background or as the centrepiece of a country picnic. Indeed, a whole industry quickly developed around the idea of countryside touring, from the introduction of the 'touring car' to the publication of guidebooks for the motorist. Picnicking and even camping equipment specially designed for use with the car soon followed. In the USA the automobile was enthusiastically adopted by the wilderness camping and touring fraternity. By the 1920s 'motor-camps' and 'auto-road camps' were catering to the needs of the motoring tourist. This dramatically altered both the pattern and extent of national park use, 'changing the wilderness from a gallery reserved for a discriminating few to a playground where all might absorb what they could' (Schmitt 1969: 155). Although the railroads were still officially regarded as the primary mode of travel to the national parks, in 1917 more than 55,000 automobiles were registered as having visited the parks. By 1926, this had risen to over 400,000 (Schmitt: 161), clear evidence of the rapid growth in the use of the automobile for recreational travel.

Although long-distance touring had become popular by the 1920s, it was the countryside surrounding urban areas which was most affected by the recreational use of the automobile. The twenties were expansive years. The middle class grew rapidly in both size and affluence. Leisure, especially the two-day weekend and the paid vacation became a permanent feature of even working-class society. And, thanks to Henry Ford, the automobile was now a possession to which ordinary folk could aspire. All the ingredients for the casual enjoyment of the countryside on a large scale were in place. Day and even afternoon trips into the country became a popular pastime, for they no longer required the planning of pre-automobile days. The pressure on scenic spots and recreational sites increased rapidly. In the USA this led to the rapid expansion of park and forest reserve systems around major cities to accommodate the demand for outdoor recreation in the countryside. By 1928, for example, New York City residents could travel to 655,000 acres of forest reserves within a 200-mile radius (Schmitt 1969). They could also travel to these areas on a growing network of parkways, specifically designed to take the motorist out into the countryside (Newton 1971).

In Britain car ownership spread far more slowly than in North America. Yet, the many books and guides to the countryside which appeared during the Edwardian and the inter-war years were oriented to the automobile. The joys of motoring for pleasure were widely publicised in books like Geoffrey Farnon's *The Open Road* (1910) and in the establishment of automobile clubs. The Royal Automobile Club was formed in 1909, followed a few years later by the Automobile Association. Together these

Plate 4.3   The joys of the open road, 1925

Plate 4.4 'See America First!' Traffic jam in Yosemite National Park, 1912

offered not only road service to members but also actively promoted with travel guides, maps and hotel endorsements the practice of automobile touring. Despite the lobbying of automobile and amenity organisations there was little attempt to provide new roads or recreational areas for the growing numbers who were driving into the countryside on weekends and holidays. And so began the now familiar congestion of picnic sites like Epping Forest, of scenic areas such as the Cotswolds and of hiking and climbing country such as the Peak District. Furthermore, the sheer density of rural settlement and its associated road network encouraged the habit of just driving around the countryside as a pastime. Villages and country lanes became inundated with cars on summer Sundays in an early preview of the mass invasion of the countryside by car-bound tourists.

The post-war boom

The inter-war years were the culmination of a century of radical changes in the recreational use of the countryside. From being the exclusive preserve of the landed, leisured classes and a few adventurous tourists in the eighteenth century, countryside recreation by the end of the 1930s had become a truly mass activity. Indeed it would be fair to say that much of the character of the modern use and perception of the countryside as a recreational amenity had been established. Four basic trends had emerged. Firstly, outdoor recreation,with its emphasis on the physical enjoyment of nature and open space was now a popular leisure activity. Secondly, significant areas of countryside had been set aside by both public and private agencies to satisfy and manage the demand for this type of recreation. This was particularly evident in the expansion of the parks system in North America. Thirdly, through the automobile, people had discovered the joys of touring. Finally and most significantly for the subsequent development of countryside recreation, the weekend and the automobile had emerged as inseparable institutions and opened up rural areas around cities for casual relaxation on a large scale.

The boom in countryside recreation that has occurred in the past 30 years must be seen, then, as a continuation of pre-war trends. The Second World War itself, of course, interrupted the normal patterns of travel and leisure, and, in Britain in particular, it was not until the 1950s that circumstances favoured their return. However, since then, social and economic conditions have become conducive to the use of the countryside as a recreational amenity on an unprecedented scale. While the factors involved were not new – rising affluence, increased automobile ownership, improved roads and more leisure – the magnitude and rate of their change was. Real incomes increased more rapidly for more people than ever before, car ownership (at least in the USA) became almost universal, highway systems were radically modernised, and leisure time steadily expanded. Within these developments we can discern the link between the two main strands of modern countryside recreation: mobility and leisure.

Universal leisure is an entirely post-war phenomenon. Although, as we have seen, there was a steady expansion of general leisure time during the twenties and thirties, it was not until the fifties that paid vacations and the two-day weekend could be enjoyed by all working groups (Rubinstein and Speakman 1969). In recent years the movement to increase the length of vacations, to shorten the working day and week, and to extend the period around public holidays has gathered momentum. With the notable exception of the growing army of the working poor (who usually have to work long hours to achieve an adequate income), and some professionals and self-employed individuals, most people now have an estimated average of twenty-five to fifty hours of 'spare' time a week; more than is consumed in paid work (Kraus 1984). In addition, increases in life expectancy, combined with significant improvements in pensions has produced a new generation of 'senior citizens' with time and (for some) money to enjoy virtually unlimited leisure.

That this newly affluent, mobile and leisure-oriented society was using the countryside and the wilderness outdoors as a recreational amenity on an unprecedented and increasing scale, was already quantitatively apparent by the early 1960s. In the USA, a study by the Outdoor Recreation Resources Review Commission (ORRRC), arguably the most extensive of its kind ever carried out, sampled the recreational behaviour of 16,000 adults in 1960 and found that approximately 90 per cent engaged in some form of outdoor recreation activity (USA 1962). While some of this was urban and home based, the survey showed that the most popular activities used rural and wilderness resources, and estimated that over half the American population took at least one trip into the countryside during the year. Furthermore, over 20 per cent of the average American's weekends and public holidays were spent in non-urban outdoor recreation. The ORRRC study predicted the continued and rapid expansion of outdoor recreation. In fact its projections fell considerably short of the actual growth recorded in subsequent surveys. The Bureau of Outdoor Recreation Survey, carried out in 1965, recorded an overall increase in participation in outdoor recreation of 51 per cent from 1960 (Simmons 1975). Even allowing for problems of survey comparability this represents a substantial increase over a short period. By 1977 the latest national survey, this time by the Heritage Conservation and Recreation Service, revealed not only that participation in the most popular outdoor activities had risen to include well over 50 per cent of the population, but that almost 60 per cent of those surveyed valued outdoor recreation more highly than other pastimes (in Cordell and Hendee 1982).

It was not until the Countryside Commission began systematic surveying of countryside recreation in the mid–1970s that any comparable trends can be recognised in Britain. A few studies, all less comprehensive than those carried out in North America at the time, did provide a rather sketchy picture of outdoor and countryside recreation participation levels during the 1960s. The Pilot National Recreation Survey, from a small sample of just over 3,000 households in the late summer of 1964, indicated that 34 per cent of respondents reported visiting the countryside on their last day or half-day outing (British Travel Authority 1967). In another study, this time of recreational travel in central Scotland carried out in 1969, 31 per cent of respondents reported visiting the countryside on their last outing, while 22 per cent reported visiting a beauty spot. This suggests that about half the population used the countryside as a recreational amenity on at least one occasion during the year (Coppock and Duffield 1975).

The clearest evidence of a general increase in the level of participation in countryside recreation, however, comes from the surveys carried out by the Countryside Commission for England and Wales. The latest of these, completed in 1984, asked people how many day or part-day trips they had made to the countryside in the previous four weeks during periods of winter, summer and spring (Countryside Commission 1985a). Over half of those interviewed had visited the countryside for recreational purposes at least once during the winter, 60 per cent had done so in the spring, 70 per cent in the summer months of July and August and 84 per cent reported visiting the countryside at least once during the year as a whole. While differences in survey methodology prevent full comparison with the studies carried out in the sixties, this does imply a substantial increase over the twenty years in the proportion of people using the countryside for recreation on at least an occasional basis. More accurate trends can be discerned from the Countryside Commission's own surveys carried out in 1977 and 1980 which are directly comparable with the 1984 study. Despite a short-term decline between 1977 and 1980, the overall trend appears to be that of a continued growth in the level of participation in countryside recreation (Countryside Commission 1985a). It is worth noting, however, that approximately 30 per cent of the population never visited the country-side at all. Furthermore the overall participation rate, as in most previous surveys on both sides of the Atlantic, is based upon the frequency of minimal use and therefore tends to give an exaggerated impression of the general popularity of countryside recreation. However, the 1984 data shows that 38 per cent of respondents in the British survey were frequent users of the countryside especially during the summer. This represents an influx of several millions into the countryside every summer weekend. Measured in these terms it has become, according to the Countryside Commission, by far the most important form of outdoor recreation in England and Wales (Countryside Commission 1985a).

THE NATURE OF COUNTRYSIDE RECREATION

What do all these people do when they visit the countryside and in what ways is it viewed as a recreational amenity today? To some extent they do much as they have done for most of this century except, as we have seen, in much greater numbers. However, the increasing affluence, mobility and leisure time of recent years has created the conditions for a broad diversification and commercialisation of countryside recreation. Furthermore, thanks to the seemingly incurable habit for statistical surveying that dominates recreational research today, we also know a great deal more about the range of activities that are pursued in the countryside. Much of this research is preoccupied with dichotomising outdoor and countryside recreation into two groups of activities: the casual, passive and informal as opposed to the organised, active and formal.

While this may be a reasonable basis for classifying the recreational activities themselves, it provides only a limited picture of the kinds of recreational experience people seek from the countryside, and therefore of the general perception of its role as a recreational amenity. A more relevant approach for our purposes, therefore, is one which recognises the values and objectives of the participants. There is, I suggest, a more profound dichotomy in countryside recreation than that outlined above. It is one which emphasises the distinction between those who view the countryside primarily in terms of the space and resources it can provide for particular kinds of outdoor activity and those for whom the love of countryside itself is the reason for its recreational use. This has been characterised by some as a simple distinction between consumptive and appreciative activities (Jackson 1986; Lavery 1971). Although this may appear to be just another simplistic dichotomy, it does, in fact, reflect the historical realities of recreation in the countryside and has determined much of its character as a recreational amenity. One thrust has indeed cast the countryside in the role of a recreational facility containing a set of resources which can be consumed by a variety of activity-centred outdoor pursuits. The other, however, has reflected a quite different perspective, one in which the recreational activities derive directly from the appreciation of the countryside and which are very much an expression of the countryside ideal which is the theme of this book.

The countryside as a recreational facility

Long before the urban masses began to look to it for recreational relief, and even before the country gentry took an interest in its pastoral virtues, the countryside was valued as a sporting facility. The traditional rural sports of hunting, shooting and fishing are, of course, ancient pursuits. As the sport of royalty and nobility, hunting is the earliest example of an activity which consumed large areas of countryside for its singular use, through the establishment of hunting parks and forests. And once the country house ideal took hold in Britain, even more land was used for the sporting diversions of the country gentleman. Because of their control of land for these purposes, it was also virtually the only active recreational use of the countryside.

Together, hunting, shooting and fishing remain the most popular countryside sports on both sides of the Atlantic. In Britain there is still a good deal of foundation to the image of hunting and shooting as the preserve of the landed classes. The 'glorious twelfth' is still a major event in the calendar of anybody who is anybody. And, while the game preserves and the grouse moors may have shrunk in number and size, and the old, disgraceful game laws may have gone, large landowners can still guarantee a weekend's sport in the season. And, despite the conventional image of its exclusivity and the recent animal rights campaigns, hunting in general in Britain attracts a large following. It is safe to assume that much of this activity still involves those with some connections to the landed interests of the countryside, and in particular the new wave of country property inhabitants. Indeed, one only has to peruse the pages of *Country Life* and *Field and Stream* to recognise the major source of participation in these pursuits and, what is more, to understand the extent to which the country-side is viewed as a sporting resource by the landed classes.

Angling in Britain has long been a popular activity with all classes, and even though some of the best fishing rivers are still located in private game preserves, many people without rural property now visit the countryside for a day's fishing. In the USA fishing is the most popular of all outdoor sporting activities (Cordell and Hendee 1982). Indeed in North America in general there is now an industry devoted to wildlife sports. Hunting has never been the exclusive activity that it has tended to be in Britain. In fact, hunting in North America, except in a few examples of snobbish imitation of the British fox-hunt, does not mean riding to hounds, nor does it have the same class connotations of British game shooting. The average North American hunter and fisherman is more likely to be a blue-collar worker.

Hunting for sport (as opposed to the traditional hunting of native peoples) in the North American context consumes during the season huge areas of countryside, but mainly its remoter, largely unsettled wilderness regions. As an activity which is dominated by the stalking and shooting of all manner of innocent game from moose to rabbits, and one which reputedly involves over millions of people in Canada and the USA each year, it is surrounded by a sophisticated paraphernalia of consumerism: the latest boats, outboard motors, all-terrain vehicles, guns, camouflaged clothing, and so on. In 1980 an estimated $39 billion was spent in the USA on wildlife recreation equipment (Cordell and Hendee 1982). What more explicit illustration of a consumptive use of the countryside could there be?

Although they continue to be popular, traditional rural sports are now only one of many activity-centred recreational pursuits which make exten-sive use of the countryside. Outdoor recreation has not escaped the impact of the increasing ingenuity with which western society uses its leisure time.

One aspect of this has been the huge growth of interest in sporting pursuits in general. For the majority of the population this is usually limited to a spectator role now that professional sports has become a massive entertainment industry in its own right. But, although it involves only a relatively small proportion of the population, there has been a rapid growth in recent years in the level of active participation in a surprisingly wide range of sporting activities.

In all of these activities the countryside is viewed as a physical resource in four basic ways. Firstly there are what can be termed terrain-based activities, like skiing and rock-climbing, which demand particular topographic environments. Ski resorts are a good example of the locational impact of sporting activities on the countryside, for it is because of their specific terrain needs that they have transformed the character of those areas which happen to contain the topographic features that they require. Secondly, there is that very popular group of activities which depend upon water resources and make extensive use of rivers, canals, lakes and coastal areas. Like skiing these have had a widespread impact on those areas of the countryside which happen to contain the best facilities for water sports. One excellent example of this is the rapid growth in Britain in the popularity of sailing which has resulted not only in the development of coastal marinas but also in the establishment of inland sailing centres on water bodies created out of abandoned gravel pits and often built into the facilities of country parks. Another good illustration is the long standing conversion of the literally thousands of lakes on the more southerly parts of the Canadian Shield in Ontario to what are effectively private water sport resources.

Yet another group of activities, however, demands not so much particular environmental features from the countryside, but rather simply the open space, the land which only the countryside can provide. Perhaps the best example of these are golfing and horse-riding, two of the more enduring and popular outdoor pastimes. Golf is probably the most extensive user of land for sporting purposes, for the average eighteen-hole course requires about 100 hectares. With the proliferation of courses, especially on the fringes of urban areas this has resulted in the conversion of quite large areas of land to a single sporting use. By the same token, horse-riding depends upon both space for stabling and riding and land resources for feeding horses. It is symbolic of the changing role of the countryside that the horse has been transformed from an essentially utilitarian to an almost entirely recreational animal. From being a largely upper-class pursuit, riding is now enjoyed by people from a wide range of backgrounds. Rural areas on both sides of the Atlantic are now dotted with stables, equestrian centres and competition grounds, as well as being criss-crossed with riding trails.

These trails, even the traditional bridle-paths which are such an historic feature of the English countryside, are rarely dedicated to the exclusive use of horse-riders. Consequently those who ride them face growing competition from other users, from the gentle hiker to the noisome trail-bike rider. While hikers and horse-riders can generally co-exist, the recent boom in the use of off-road motorised pleasure vehicles represents a serious threat to both groups. There can be no more explicit example of an intrusive use of the countryside for recreation than this new craze for all-terrain vehicles. While it is mainly limited to motor bikes in Britain (some of which are used in organised moto-cross events), in North America it involves an apparently increasing range of equipment: snowmobiles, a variety of four-wheel drive vehicles and the ubiquitous trail-bike. Recent figures estimate that there are over 1.5 million registered snowmobiles and almost the same number of off-road motor bikes in the USA (Cordell and Hendee 1982).

Although it is mainly through the sporting activities described above that the countryside is treated as a recreational facility, those who pursue less active and less intentioned forms of outdoor recreation often cast the countryside in the same role. The historical tendency for urban dwellers to see the open space surrounding cities as their outdoor playground has generated persistent demand for areas where people can go for a family picnic, a leisurely stroll, to swim and sunbathe, and even play a few casual ball games. Once parks became part of the urban fabric much of this demand could be satisfied closer to home, yet the near-urban countryside continues to attract those looking for more informal and spacious environments. Indeed, so great has been the demand for leisure facilities in the countryside, that governments have responded by designating parks for intensive recreational use in order to relieve the pressure on natural and cultural countryside environments. I have already described one of the earliest versions of these in the parkway and country park systems instituted in the New York region in the 1920s. User-oriented parks, with provision for car-parking and a range of facilities for picnicking, swimming, walking, playing games, even boating and fishing, are now a common feature of many metropolitan regions.

The appreciation of countryside

In some ways the distinction between consumptive and appreciative recreational uses of the countryside is a subtle one. It is a fine line which separates the hiker who enjoys the exercise and the sense of adventure but coincidentally appreciates nature and scenery, from the individual who hikes in order to enjoy these same features. Indeed some hikers may not be able to recognise the distinction in their own recreational motives. Even the angler whose main concern may be the sport itself need not be ignorant of the more pastoral pleasures of the river bank.

There can no better illustration of this compromise of purpose than today's outdoors enthusiasts, who, in direct line from the turn-of-the-century nature and outdoors movements, combine the enjoyment of scenery and nature with exercise and exploration. This is one of the fastest growing areas of outdoor recreation of recent years, especially in North America where the wilderness and backwoods tradition remains strong. Hiking and backpacking, together with back-country camping and canoe-tripping are the most popular wilderness activities. In Yosemite National Park, for example, participation in back-country camping increased from 78,000 to 221,000 between 1967 and 1972 (Hammitt and Cole 1987), while between 1965 and 1980 visits to the National Wilderness Preservation System more than doubled (Cordell and Hendee 1982). In Canada, wilderness tripping has grown to the point that it involves close to a quarter of those who participate in outdoor recreation. (Jackson 1986). In both countries a whole industry has grown up around the wilderness experience, from the sale of equipment to the recent boom in commercially organised wilderness 'adventures'. Jackson draws a clear distinction between consumptive and appreciative wilderness recreation, but in truth, for the hiker, backpacker and canoer the activity itself is not so easily separated from the surroundings. For some, in fact, the true experience of wilderness and nature can be achieved only through the direct contact which comes with the activities themselves. The attraction is in the challenge of survival and the motivation is the escape from civilisation, communion with nature, even perhaps a hint of transcendentalism.

The experience of the outdoors enthusiast in Britain is considerably more muted and largely unconnected with the true wilderness experience. Yet the long tradition of hiking and rambling continues today, representing, as recent surveys suggest, an estimated 20 per cent of recreational activities in the countryside. Its sustained popularity is reflected too in the sizeable memberships of the Ramblers' Association and the Youth Hostels Association (see Chapter 6). The hikers' 'wildernesses' are the uplands of Britain: the fells of the Lake District, the limestone and gritstone moorlands of the Pennines, the wilder reaches of the Scottish Highlands and the Welsh Mountains. National parks (at least in England and Wales) and a growing network of long distance footpaths cater to the needs of of the serious explorer of these remoter stretches of country as the map in Plate 4.5 shows. There is now an extensive network of long distance footpaths in Britain. However, both parks and footpaths also encompass and traverse the country's tamer and more settled landscapes. Together with the extensive network of local foot- and bridle-paths they are possibly the most valued recreational resource of the British countryside, especially now that their appreciative use is threatened by off-road motor cycles and other such vehicles. That the combination of physical exercise and countryside appreciation may be growing in popularity is evident not only in the

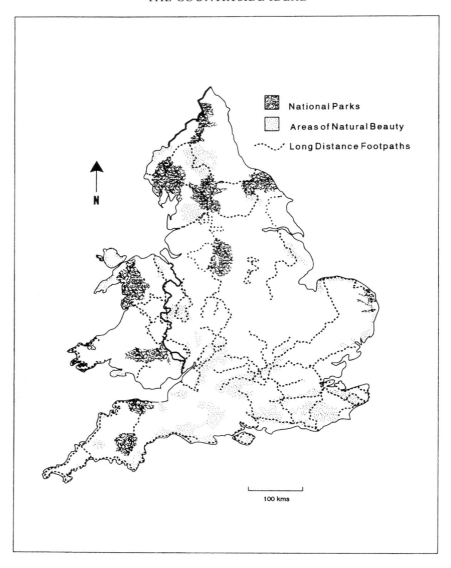

Plate 4.5 On every city's doorstep: National Parks, Areas of Outstanding Natural Beauty and long-distance footpaths in England and Wales, 1992, © BTA 1993, Licence No. 123

expansion of the footpath network but also in the introduction of so-called 'activity holidays'. The Countrywide Holidays Association now offers 'walking holidays of various grades based in national parks, also rock-scrambling, pony-trekking and ... adventure and activity holidays for

130

children and teenagers', while Ramblers Holidays Limited advertises 'walking and trekking holidays accompanied by a leader who plans and leads each day's activities'. Horse-riding has long been a popular way to explore the countryside and to enjoy the wilder areas of the country, and pony-trekking has certainly grown in popularity in the past few years. In addition cycling, camping and caravanning have all become part of a widening interest in the direct experience of nature and countryside. Campsites and caravan or trailer parks are scattered all over the countryside on both sides of the Atlantic, offering cheap holiday accommodation yet at the same time often having an intrusive impact on the landscape as Plate 4.6 illustrates.

Plate 4.6   Regimented recreational intrusion into the rural landscape of the Isle of Wight

It would, of course, be wrong to give the impression that the appreciation of the countryside necessarily demands the spirit of adventure and the level of physical exercise which I have just described. When the Victorian working classes went on to the moors and into the dales for a few precious hours of Sunday walking, they sought casual and generally sedate relief in nature, open space and scenery. Just 'getting out into the countryside' became an end in itself and the essential ingredient of its mass recreational use. Today, of course, the car has become virtually the universal basis of this kind of countryside appreciation. In addition to improving access to particular recreational facilities and resources, it is also a recreational experience in its own right. Driving for pleasure and sightseeing remain the most popular North American modes of countryside appreciation. In Britain the car has become the focus of countryside recreation. 'Just touring around', 'just out for a drive', 'a trip out' were the most commonly reported activities in a Scottish survey in the late 1960s (Duffield and Owen 1970). Nearly fifteen years later, the Countryside Commission revealed that going for a drive was still the most popular way of visiting the countryside (Countryside Commission 1985a).

Although it may not be aimed at any particular destination, motorists can focus their appreciation of the countryside on a wide range of sites and attractions. Especially reflective of the traditional affection for countryside as landscape and nature is the trip to a natural beauty spot where the car can be parked and the scene enjoyed. In the North American context this invariably means a publicly-owned park where a variety of other outdoor recreation activities are pursued, for, apart from roadside pull-ins or 'viewpoints' (as highway departments often insist on calling them), the settled countryside is largely inaccessible to the casual visitor. By contrast the British countryside is dotted with small patches of land which provide attractive stopping places for the weekend motorist, ranging from common land and areas preserved by the National Trust, to country parks and the properties of the Forestry Commission. The National Trust, as we shall see in more detail in Chapter 6, has taken a particularly active role in preserving sites for casual public appreciation, and its properties are veritable 'honeypots' for motorists on a fine weekend. The car, however, remains very much the centre of recreational activity, for, even when they have reached their destination and parked, the majority seem to limit their appreciation of the natural scene to the area around the car park. The impression that the trip into the country is a largely sedentary form of appreciation is borne out by Burton's study of Cannock Chase, a popular country park in the English Midlands, where she found that most visitors stayed either in or by their cars and, of those who did go for a walk, few went more than 300 metres from the parking area (Burton 1973).

If the sedentary enjoyment of landscape and scenery fails to satisfy, there are plenty of other 'country' attractions from which to choose. In Britain traditional destinations on a country drive are stately homes, picturesque villages and places of historic or literary significance like the Lake District or Hardy country (Plate 4.7). Overseas tourists are now fed a predictable English countryside diet as they traipse from Wessex to the Cotswolds, stopping to admire the picturesque perfection of Castle Combe and to sample the tea 'shoppes' of Broadway, before heading on to Blenheim or Chatsworth and thence to the literary nostalgia of Wordsworth country in the Lake District. Essential ingredients, too, are the country pubs and the antique and craft shops which complete the image of a Britain which is still rural at heart. Not that this simplistic and largely commercialised

Plate 4.7 Pilgrimage to Wessex: Thomas Hardy's cottage at Higher Bockhampton, Dorset

nostalgia goes unappreciated by the the British themselves for they too converge on these symbols of a bygone age aided by their AA and Shell guides to the countryside. There seems to be a special interest in recent years in the reconstructed rural past, especially in the growing number of country parks which are heavy on pre-industrial imagery and agrarian nostalgia. A typical example of this is Bickleigh Farms in Devon, a privately-owned park established to tap the local holiday trade (families looking for something to do as a refuge from a wet and chilly beach). Complete with working replica of a water-mill, potters, basket-makers and weavers, old farm implements and docile livestock, a farm shop purveying suitably wholesome local fare, and an outlet for country-style tweeds and woollens, Bickleigh is, in its own words, the 'complete country experience' (Plate 4.8).

This is a standard formula in countryside tourism, especially in North America where the appreciation of countryside as a recreational activity is strongly influenced by nostalgia for country life. The experience of bygone rural days in pioneer villages, 'antiquefied' small-town main streets and country stores seems to have become a particularly popular diversion for those seeking suitably 'country' things to do (and consume) on a country drive. While some of this has been debased to the amusement park category, most of this commercialised countryside has developed to serve the affluent consumer, for whom a weekend staying at a classy country inn and browsing around antique and craft shops has become very much an extension of fashionable lifestyle. The commodity which seems to be most sought after is 'country atmosphere', and villages and small towns in most of the metropolitan hinterlands of Canada and the USA have been quick to re-create this atmosphere in their enthusiasm to capitalise on its commercial potential. A good example of this is the Amish region around Lancaster in southern Pennsylvania where a whole economy has been converted to the peddling of rural nostalgia. Not all of this activity is influenced by commercial or consumerist motives. In the restoration of pioneer villages and of the architectural fabric of surviving rural places, there is, of course, a serious historical interest, one which promotes the appreciation and knowledge of rural heritage. Old Sturbridge Village in Massachusetts is probably one of the most visited of such sites in the USA. With its attention to the details of the domestic and agricultural lifestyle of the times, it is typical of the educational yet nostalgic tone of the many restored colonial and other pioneer villages across the continent. At Sturbridge the central theme is authenticity, from building construction to soap-making and in special events such as sheep-shearing and harvest festivals (Plate 4.9).

Old Sturbridge sets out to foster an understanding of rural history and heritage; to encourage, that is, an interest in the rural past which, however nostalgic, is more academic than commercial. In this sense it is part of a

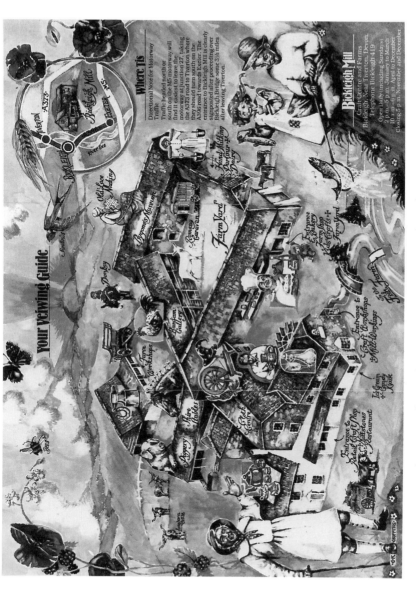

Plate 4.8 'The complete country experience' at Bickleigh Mill, Devon

Plate 4.9 The educational countryside: pioneer hay-making at Old Sturbridge, Massachusetts

broad tradition in which the appreciation of nature and the outdoors, and of the intricacies of the countryside in general are the focus of the recreational experience. The most popular and enduring strand in this tradition is the study of nature. Amateur naturalism, which played such a central role in recreational use of the countryside by the leisured classes in the early nineteenth century, and which had grown into a popular pastime by the end of the nineteenth century appears to have experienced a resurgence of interest in recent years. Not surprisingly there is a close relationship between this kind of outdoor recreation and environmental and countryside conservation movements. These are discussed in detail in Chapter 6. However, it is worth noting here that it is the public agencies responsible for conservation which have been at the forefront of the promotion of the educational aspects of outdoor recreation. This is consistent with the growing argument for the encouragement of more constructive and educational uses of leisure time. In the USA both the National Parks and the national Forest Services have long provided interpretive opportunities for visitors, while the more recently established Wilderness Preservation System fosters the appreciation of wildlife and natural environment in the specific context of wildlife and nature reserves. It is in these settings that the tension between recreational and environmental conservation is most evident. Yet it is for this very reason that the agencies concerned place so much emphasis on the educational component of park use.

In Britain, with its long history of protectionist concern for the countryside, the study of nature, indeed countryside interpretation in general, has been carried to a fine art. It is impossible to estimate the number of amateur naturalists in Britain today, but there are a plethora of organisations which play an active role in promoting the serious appreciation as well as the protection of the natural environment of the countryside. At the government level the Nature Conservancy Council oversees a national system of nature reserves, and, in addition to its primary mandate of nature conservation, plays a central role in educating the public in the wise use of the countryside. There is in all of this a strongly élitist thrust, but in recent years there has been a trend towards spreading the gospel of nature study to the public at large and in particular towards its integration into the recreational use of the countryside. This is due in no small measure to the growing concern for the wise management of recreation in the countryside.

The main objectives of public agencies such as the Countryside Commission and of the many private conservation groups are to foster greater respect for and understanding of the countryside, to increase awareness of its amenity value and to promote environmental and heritage conservation. Countryside interpretation seems to be experiencing a surge of popularity in Britain today, both among active participants and armchair enthusiasts. For the latter, as we have already seen in Chapter 2, there is a steady stream of publications on all manner of countryside matters, which for some presumably whets the appetite for the on-site appreciation of nature, landscape and country ways. This has been matched by an impressive expansion of resources to provide visitors to the countryside with interpretive opportunities. These include information centres, nature trails and forest walks, exhibitions, pamphlets, wayside signs, audio-visual displays, guided walks and farm open days. In the New Forest, the Rhinefield Arboretum provides attractive walks along paths which wind between the many varieties of trees and shrubs which make up the site. Interpretive plaques about the various species as well as forest ecology line the route. It is a popular spot for an afternoon outing; one of several places on the New Forest recreational itinerary. My own casual observation of its use on a summer Sunday afternoon suggests that most people take only passing interest in the interpretive information and that only a few families (often with a parental zeal which produces yawns from their offspring) get seriously involved in the educational programme.

The Arboretum is a good example of the way in which most visitors to the countryside respond to the interpretive approach to recreation: casual interest rather than passionate involvement. The use of country parks is a case in point. Particularly in local authority managed parks but also in some privately run parks, nature and country life interpretation is an integral part of the recreational facilities. Yet, if the Cannock Chase Country Park study is any indication, few visitors do little more than take a short stroll. This is not to say that visitors to country parks are indifferent to their interpretive programmes. After all some parks now devote a large part of their budgets to nature study and appreciation. This is consistent with a more general movement to promote countryside interpretation as a recreational activity in its own right. Organisations like the British Trust for Conservation Volunteers and the National Trust offer 'conservation working holidays', while the local National Park Societies organise guided walks, lectures and other educational programmes as part of their mandate to garner support for the protection of park environments. The British Trust, for example, now offers countryside skills courses, which include dry-stone walling, hedgelaying, tool repair and fencing skills.

Not surprisingly, the Countryside Commission has been at the centre of interpretive initiatives. Its Groundwork programme organises volunteer involvement in the improvement of the countrysides of urban fringes. Tree planting, footpath improvement and waymarking, installing stiles and gates and removing rubbish are typical volunteer activities, while nature walks form the basis of the educational component. With the encouragement and often the direct involvement of the Countryside Commission several programmes are directed at schoolchildren. A number of 'holiday play-scheme projects', based on country parks and designed to provide children with countryside experiences during the school holidays have been initiated. Irchester Country Park, near Wellingborough has its 'Operation Woodpecker', a countryside playscheme for 8 to 14 year olds. At Waseley Hills Country Park, south-west of Birmingham, 'Project Countryside' offers children the chance to participate in pond-dipping, bird-watching, nest-box making, conservation games and quizzes, arts and crafts, as well as learning tracking and exploring farm trails (Countryside Commission 1985c).

Given the social and economic trends of the modern industrial age it was inevitable that the countryside should become a major recreational amenity. The simple explanation for this would be that it was provoked by the need for relief from the conditions of the urban–industrial city and of its suburban offspring. Certainly I have argued throughout this book that this is a persistent theme in the countryside tradition as a whole. In this sense the use of the countryside for recreation can be seen both as an expression of broader cultural values and sentiments, and as a fulfilment of basic physical and psychological needs. It is the 'armchair countryside' made real; the pursuit of myth and nostalgia in the landscape itself, as well as a manifestation of those fundamental needs for contact with nature, for aesthetic and sensory pleasure, for freedom and space. The growth of countryside recreation, however, must also be understood in the context of general changes in western lifestyle; in the context, that is, of an increasingly affluent, mobile and leisure-oriented society. It is a lifestyle which permits the virtually unlimited fulfilment of the values and needs described above. Yet it is also a lifestyle which requires ways of enjoying mobility and spending money and leisure time. Recreation in the countryside, whether consumptive or appreciative, has become part of that lifestyle. The countryside is a diversion; a place to go and explore as travel becomes a leisure activity in its own right, a place to consume as disposable incomes increase and a place in which to relax and play as leisure time expands.

All this has had far-reaching consequences both for our perception of the role of the countryside and for its actual character. That it was increasingly being perceived as a recreational amenity was already apparent over a century ago. Since then this perception has matured to the point that today's countryside is regarded by many as serving a purely recreational function. As such it is cast in a multi-dimensional role: a sports facility, a public park, a tourist attraction, a therapeutic resource, a nature reserve and a museum. And the character and appearance of much of the countryside has come to reflect these roles. Sporting facilities from golf-courses to race-tracks have taken over chunks of agricultural land, while outdoor pursuits from hunting to hang-gliding have invaded both the settled countryside and the forest, bush and wilderness beyond. Indeed, with aircraft and all-terrain vehicles not even the remotest areas are left untouched. As a tourist attraction the countryside is the consumption of scenery from the road, the spectacular view from the signposted vantage point, the picturesque village, the historic site, the campground, the bed and breakfast cottage and, increasingly, the commercialised and often trivialised versions of country life. Above all, it is the countryside of coaches and cars, service stations and stopping places, snack bars and tearooms, restaurants and roadside inns. As a therapeutic resource it is a countryside of hiking trails, footpaths and bridleways, accessible open spaces and preserved landscapes. And finally, the countryside is a repository of natural and cultural history, set aside in a growing number of protected areas, revealed and directed in countless nature trails, and enshrined in reconstructed villages and country museums. In many ways, then, the countryside has become a huge outdoor recreational facility; its landscapes and land uses, its economies and communities fundamentally and probably irrevocably transformed to conform to the edicts of a leisure-oriented society.

5

THE COUNTRY IN THE CITY

The prevailing theme so far in this book has been that of escape; of the countryside as a mental and physical refuge from urban life. However, countryside sentiment has not just involved a turning away from the city. It has also permeated the development of the modern urban landscape itself. Nature, rural scenery, vernacular architecture, village settlement form and the nostalgically defined idea of countryside in general have been the source of much of the inspiration for the treatment of the ills of the industrial city; for the search for more liveable urban environments and for the ideal metropolitan form. In short, a persistent Arcadian thread runs through the landscape history of the modern western metropolis. It is manifested in the development of parks and the preservation of green spaces, in the building of garden suburbs and the related idealism of the garden city movement, and, ultimately, in the planning and design conventions of modern suburbia. In recent years, it has also been reflected in the extension of 'green' environmental values to the urban environment and a revival of interest in the village as the model for sustainable urban living.

PARKS FOR THE PEOPLE

English origins

In 1635 London's Hyde Park was opened to the public. Initially reserved for the sporting pursuits of royalty and nobility, the park, with its huge expanses of pastoral open space soon became a popular place for the well-to-do to ride and take the air (and, incidentally, for thieves and highwaymen who took advantage of its wildness and seclusion). A few years later St James's Park was also opened to the general populace. In his *Tour of London*, written in 1772, Grossley obviously approved of the park. In it, he wrote, 'nature appears in all its rustic simplicity: it is a meadow, regularly intersected by canals, and with willows and poplars, without any regard to order' (in Marshall 1968: 151).

Public open space had been an integral feature of cities since the earliest urban civilisations. The opening of London's Royal Parks, however, represented the beginning of a shift away from the artificiality of the squares and the formal gardens of the Renaissance city and its classical ancestors to the informal and pastoral urban park. In part this was a predictable consequence of the rapid growth of London during the eighteenth century which undoubtedly increased the attractiveness of accessible green spaces within an otherwise congested city. Yet, as the movement for this kind of park gathered momentum during the early nineteenth century, it was increasingly influenced by the general shift in landscape tastes and by the growing social and intellectual concern over the conditions of the industrial city.

The basic model for the urban park was the landscaped park of the English country estate. Humphrey Repton, for example, collaborated with the architect John Nash in the laying out of neighbourhoods in which prestigious mansions surrounded small, landscaped gardens (Chadwick 1966). Repton's gardens at Russell Square in central London are one of the best examples of what, in a sense, represented the transfer of the picturesque ideals of the country park to the town houses of the landed gentry. These gardens, however, were locked at night and therefore were treated largely as private rather than public spaces. In the same vein, the first major park development outside the Royal Parks, John Nash's Marylebone (later Regent's) Park, which began life in 1811 as a real-estate venture of the Prince Regent, was conceived and designed as series of classy terraced villas surrounding a large, but private park (Newton 1971). It was to be in Nash's words a 'rural paradise' which offered the attraction of open Space, free air and the scenery of Nature, with the means and invitation of exercise on horseback, on foot and in Carriages . . . as allurements or motives for the wealthy part of the Public' (in Williams 1978: 205).

The general public, in fact, were not admitted to Regent's Park until 1838. However, what Nash introduced there and in his redesign of St James's Park in 1828, represented the first explicit attempts to incorporate the conventions of the landscape gardening movement into the development of the urban park. Nash was strongly influenced, first by Repton's version of the picturesque and then by the more contrived gardenesque espoused by Loudon, of which the principal feature was the creation of carefully planned 'natural' vistas: nature and scenery incorporated into the fabric of the city. Indeed, in Chadwick's view, Nash's St James's Park can be regarded as the first public park in this tradition. Before this, Chadwick has argued, there was little need to set aside public open spaces in cities (Chadwick 1966). Of course, conditions in London and the new industrial cities by the end of the eighteenth century, contrary to what Chadwick has implied, were ample justification for providing parks for the working class.

But it was not until the 1830s that the voices of Robert Southey and others who had been railing against the vices of industrial urbanism and forecasting the dire consequences of separating working people from nature, fresh air and country scenes, were translated into a wave of official inquiries into urban conditions. In 1833 the Select Committee on Public Walks heralded the parks movement by advocating the need for facilities for working-class families to enjoy fresh air in congenial surroundings on their day of rest. The opportunity to contemplate nature and discourse with their fellow citizens was seen largely in terms of its moral benefits: 'if deprived of any such resource, it is probable that their only escape from the narrow courts and alleys . . . will be the drinking shops' (in Chadwick 1966: 111). Seven years later the Select Commitee on the Health of Towns recommended public walks and open spaces as a means of ensuring 'a more healthy and moral people' (in Laurie 1979). Public health and morals, however, were not the only arguments made in favour of parks. As Laurie has pointed out, there were several other factors involved: the continuing influence of gardenesque ideas which equated the aesthetics of 'natural' landscapes with urban improvement, the beneficial impact of parks on property values, and the growing interest in the scientific study of nature.

It was through this combination of progressive forces – social, aesthetic, commercial and scientific – that the idea of the planned public urban park was born. It involved, in Newton's words, 'the transition of landscape architecture from the service of wealthy patrons to the service of the public at large' (Newton 1971: 223). The earliest versions of this kind of park were the botanical gardens which were laid out in most large cities in the 1830s. But the most influential developments were Victoria Park in London's East End and Birkenhead Park on Merseyside. Work began on Victoria Park in 1841 under the direction of James Pennethorne, initially in the best traditions of the gardenesque school. Largely because of its limited size and uninspiring setting, the park never quite fulfilled Pennethorne's ambitions, yet it was laid out with grassed areas, tree plantings, curving pathways and small lakes and continues to provide recreational space for East Londoners today. As an expression of the improving and reformist ideas of its times, however, Victoria Park was soon overshadowed by the development of Birkenhead Park. The rapid growth of Birkenhead, like so many northern industrial cities in the early decades of the century led to the creation of a Board of Improvement, the Commissioners of which soon recognised the reputation that a fine public park would acquire for the city. Yet the more socially conscious Commissioners also argued for the benefits of a 'Gentleman's Park' for the working man in which industrial conditions would be relieved by country-like scenery (Newton 1971).

Plate 5.1   A Summer's Day in Hyde Park, by John Ritchie, 1858

Birkenhead Park was laid out by Sir Joseph Paxton, who was later responsible for the Crystal Palace and the landscaping for the 1851 Great Exhibition in Hyde Park. Paxton was strongly influenced by Loudon, and so, not surprisingly, designed a park to replicate natural scenery, with lakes, shrubberies, scattered clumps of trees separated by large meadows and traversed and encircled by the usual curving pattern of pathways of the gardenesque school. In tune with the real-estate philosophy of the time the park was surrounded by terraces of mansions. It was, however, the largest public park yet built, covering some 125 acres, excluding the area set aside for houses. It was opened in 1847 and quickly became famous amongst both civic reformers and landscape architects, including the American Frederick Law Olmsted (Chadwick 1966).

Other cities followed Birkenhead's lead in developing large public parks. Again London was in the forefront with the park at Kennington in 1858 and the 100-acre park at Battersea incorporated into the development of the South Bank in 1868. And in mid-Victorian London people flocked on Sundays and holidays to the Royal Parks where the main pleasure, as the painting of Hyde Park in the 1850s shown in Plate 5.1 illustrates, was the casual enjoyment of the pastoral scene. That natural open space was becoming increasingly valued by the inhabitants of Victorian cities was also evident from the pressure to preserve the patches of common land which were in danger of becoming engulfed by urban expansion (Prince 1986). The campaign to save and to guarantee public access to common land was led by Britain's first conservation group, the Commons Preservation Society. Although the Metropolitan Commons Act 1866 provided for access as well as the management and control of urban commons, it took vigorous campaigns such as that for the preservation of Hampstead Heath and Wimbledon Common to ensure the maintenance of their original size and landscape (Laurie 1979). In several cities, in fact, commons were often the only remaining sizeable undeveloped areas and so, as the demand for informal recreational space increased, they became *de facto* public parks. Southampton Common is a particularly good example of this. Originally a Saxon common, it became part of the Town of Southampton in 1228 and has been held in trust for its citizens ever since (Laurie 1979). The city council now acts as trustees for a park of 146 acres which still retains much of its woodland character despite the introduction of sports fields and children's playgrounds (Plate 5.2).

The Olmstedian tradition

Although it was an idea that originated in Victorian Britain, it was in the North American city in the second half of the nineteenth century that the large pastoral urban park achieved its full potential. The chief protagonists for American parks gained their initial inspiration directly from English landscape gardening principles in general and from the extension of these principles to early English urban parks in particular. Andrew Jackson Downing, whose voice was among the first to be raised in support of natural open spaces in American cities, was directly influenced by his visits to English country estates and London's Royal Parks. And Frederick Law Olmsted, the most influential figure in the history of urban parks, gained

Plate 5.2   Nature in the city: Southampton Common, 1993

many of his early ideas from visits to England, where, as we have seen, he was profoundly impressed by the appearance of Birkenhead Park, but where he also acquired the sense of the social benefits of providing natural open space in cities.

However, the American parks movement must also be seen as an expression of American romanticism and its particular philosophy of nature and rural scenery. Urban parks were a logical extension of that philosophy. Among the earliest American expressions of the value of nature in urban areas was the removal of cemeteries to rural locations just beyond city boundaries, partly because of the pressure of demand for urban land but also because of the belief that more natural surroundings would permit the 'absorption of deleterious gases' from the graves, and that landscape and scenery would uplift the emotions of mourners (Schuyler 1986: 61). Mount Auburn Cemetery outside Boston and Greenwood Cemetery on the edge of Brooklyn were laid out in the 1830s following picturesque principles, and quickly became popular as places for a quiet stroll or even a family picnic. For Downing these cemeteries were a model for the creation of urban public parks: he described Greenwood as 'grand, digni-fied and park-like' (in Newton 1971: 268).

Downing's was not the only voice arguing for natural open spaces in cities. Thoreau himself advocated that each community should set aside 'a park or rather a primitive forest . . . as a place that would keep the New

World *new* and preserve all the advantages of living in the country' (in Schuyler 1986). In 1844 William Cullen Bryant had already proposed a park for New York, arguing along the lines of the British social reformers for the need 'to anticipate the corrupt atmosphere generated in hot and crowded streets' (in Newton 1971: 269). In 1848 the American Medical Association echoed this concern by advocating public parks as necessities the maintenance of urban health (Schuyler 1986). But it was Downing who became the prime mover in the early campaign for parks in American cities. For him the main purpose of the urban park was to provide a contrast to and a relief from the urban landscape. In 1851 he articulated his plans for a park in central New York. Above all it was to be a place where people could enjoy 'the substantial delights of country roads and country scenery and forget for a time the rattle of the pavements and the glare of brick walls' (in Chadwick 1966: 163).

Over the next three years the land for Central Park was assembled and in 1857 the man who was to have a profound influence on American urban landscape for the rest of the century, Frederick Law Olmsted, was appointed as its first superintendent, with Calvert Vaux as architect. At the heart of the 'Greensward' plan which Olmsted and Vaux prepared for Central Park, was Olmsted's belief that a park should be a rural landscape within the city (Olmsted 1870). Hence the emphasis on separation from the city blocks which surrounded it; on the inward-looking perspective of the park with its scenic effects based on large meadows and small clumps of woodland, which still stands out so clearly today (Plate 5.3). To this Olmsted added the ideas of the picturesque with which he had been so impressed in his visit to Birkenhead Park: the Ramble with its rocks and foliage, the ornate bridges which separated people from traffic, the network of footpaths and the lake effect in the large reservoir.

Central Park took ten years and a good deal of political acrimony to complete yet even while they were still working on it, Olmsted and Vaux were involved in projects in a growing number of cities. Olmsted prepared designs for Golden Gate Park in San Francisco, and in 1866 he and Vaux embarked on their second great project, Prospect Park in Brooklyn. Within a few years the partnership had laid out what is still the largest urban park in North America, Fairmount Park in Philadelphia, followed by parks in Albany, Newark, Providence and Hartford. In Chicago in 1869 they collaborated with Horace William Cleveland, who had been involved in early ideas for Central Park. Cleveland had begun to think of interconnected systems of parks, an idea which he first proposed in his plan for Minneapolis–St Paul and which became the basis for the much later development of the Twin Cities parks system. The Chicago plan called for two large parks joined by a parkway. Olmsted and Vaux were awarded the contract for the South Parks (today's waterfront Jacksons Park) and in its design they continued the theme of parks as naturalistic landscapes. The Chicago plan

Plate 5.3 Central Park today still stands out in sharp naturalistic relief from the rest of Manhattan

highlighted the growing debate over whether parks should be naturalistic or associational landscapes, for in his design for the Western Parks, William Jenny conceived of public open space primarily as an adjunct to buildings (Schuyler 1986).

Yet for the next twenty years it was Olmsted's concept of the park which prevailed. In writing in 1870 of the plans for Boston's Franklin Park, Olmsted drew a clear distinction between the public garden and the public park. In the garden it was urban beauty, through ornamental effects, which was achieved. In the park, the objective was the beauty of rural scenery obtained through the careful arrangement of natural elements (Olmsted 1870). Olmsted was under no illusions as to the limitations of achieving this rural effect in the city, describing the objectives of Franklin Park as 'placing within easy reach of the city the enjoyment of such a measure as is practicable of rural scenery'. What is more significant, however is that, in his plans for Boston, Olmsted stressed a metropolitan solution which would anticipate future suburban expansion (Schuyler 1986). This was to be based on a series of peripheral parks – an 'emerald necklace' – connected by parkways; 'shaded pleasure drives' along which citizens could drive their carriages from one park to another. An earlier version of this had already been achieved in the much praised Olmsted and Vaux plan for Buffalo. In Boston the core of the system was the 'Promenade', a continous parkway from Boston Common to Franklin Park. The metropolitan system as a whole was conceived by Olmsted as a series of 'green fingers', corridors of the surrounding countryside both penetrating the city and providing access to parks on the outskirts. The full development of the system took another twenty-five years and the energy and vision of Charles Eliot, who joined the firm of Olmsted and his son Clarence in 1893 and became the leading proponent for preserving scenic beauty in metropolitan areas. With the establishment of the Metropolitan Parks Commission the Boston area parks system, which is an integral part of today's suburban landscape, was given a secure institutional framework which soon became a model for other metropolitan areas (Newton 1971).

In 1873 the partnership of Olmsted and Vaux was dissolved by mutual agreement, yet Olmsted continued to work on projects all over the continent, including Belle Isle park in Detroit and Mount Royal in Montreal, as well as devoting a good deal of his time to the development of the Boston parks system. Civic-minded politicians and private citizens in city after city embraced the idea of naturalistic parks as an essential element of a civilised urban landscape. Bringing pastoral scenery into the city became a popular goal amongst those who feared the worst from land speculators and pro-development city councils. Writing in 1888, Charles Sprague Sargent expressed this in unequivocal terms:

149

An urban park is useful in proportion as it is rural. The real, the only reason why a great park should be made, is to bring the country into the town, and make it possible for the inhabitants of crowded cities to enjoy the calm and restfulness which only a rural landscape and rural surroundings can give ... all other objects must, in a great park, be subordinated to the one central, controlling idea of rural repose, which space alone can give.

(in Schuyler 1986: 144).

The implementation of the Olmstedian vision at a metropolitan scale marked the culmination of a movement which had begun as a measure for ameliorating urban conditions but which had become a general philosophy for park planning. Yet just as this view of the urban landscape reached its zenith towards the end of the nineteenth century it began to be supplanted by a quite different philosophy of parks and of how nature fitted into the urban scene. Indeed in Britain, from as early as the 1860s the development of new urban public parks increasingly emphasised the formal civic park which, with its ornamental gardens, manicured 'Keep Off The Grass' lawns and designated recreation areas, was more an extension of the built environment than a retreat from it (Chadwick 1966). Although this reflected a shift in design philosophy it was also a function of the increasing cost of land, especially in rapidly expanding suburbia. By the end of the century North American cities were experiencing similar land shortages and a political climate which was less and less favourable to the allocation of land for large parks. But the Olmstedian park also came under attack from an increasingly organised playground movement, which opposed what it regarded as the élitist aesthetics of the naturalistic park (Schmitt 1969). Across the country most new city parks were designed to serve active recreational needs, and nature moved very much into the background. Even Olmsted's parks were increasingly adapted to this new enthusiasm for sports and fitness. Tennis courts were laid out in Brooklyn's Prospect Park and in Boston's Franklin Park while Central Park became the object of repeated proposals for recreational facilities.

Another factor which led to the decline of naturalistic urban park philosophy was the rise, during roughly the same period, of what became known as the the City Beautiful Movement. This originated in the Columbian World Exposition in Chicago in 1893 (Olmsted was involved in the landscaping of parks for the Exposition), and emphasised grand civic design. Instead of being a naturalistic alternative to the built environment, parks and open spaces were to be an integral part of it. They were littered with statues and civic buildings, while nature was reduced to ornamental status in grand, tree-lined boulevards and public gardens designed to provide perspective and decoration for monumental neo-classical architecture. It was the work of Burnham in Chicago which provided the model for a movement in which even the parkway became a device for accentuating the grandeur of the built environment. It was the forerunner of the urban master plan and was a philosophy of civic development which spread to many North American cities between the 1890s and 1920s (Relph 1988).

Metropolitan parks

The early years of this century, then, were marked by considerable professional turmoil over the character of public parks and the role of nature in urban design. The attitudes of civic politicians were an important factor in this for as the pressure on urban land grew in the early years of the twentieth century, city governments not only leaned more towards the playground and public garden park models, but also were increasingly reluctant to allocate the space necessary for the large, informal park. Improvements in transportation, as we have seen, by permitting suburban and exurban development as well as providing greater recreational access to the countryside, further weakened the argument for bringing the country into the city by way of the park. For all these reasons few new city parks along naturalistic lines were developed on either side of the Atlantic after 1900. The idea of the urban public park as an Arcadian escape was very much a product of its age; of the lingering romanticism and anti-urbanism of the second half of the nineteenth century. These ideals had no place in the civic philosophy of the new century. The romantic belief in bringing the country into the city, or, as Garrett Eckbo (1969) has put it, 'of forcing nature on architecture' was supplanted by the design-oriented principles of modern landscape architecture in which nature in parks was reduced to the status of horticultural decoration.

The development of the naturalistic park, however, did not cease entirely. While it was being rapidly abandoned in the city proper, it was taking on a new lease of life in suburbia and exurban areas beyond. Much of the impetus for this came from the development of electric tram (streetcar) lines. 'Trolley parks' were frequently developed near tramcar termini and by the end of the century had become a leading attraction on holidays and Sundays (Hall 1977). The metropolitan model that originated in Boston spread to a number of other North American urban areas. Charles Eliot became a central figure in the growing environmental conservation movement, and the protection of natural open space in expanding metropolitan areas surfaced as an important issue. It was a logical extension of Olmsted's philosophy; Olmsted himself had already begun to pay more attention to the design of suburban landscapes by the 1880s. By 1905 the Boston system had expanded to 6,000 hectares of city parks and peripheral reservations all within a radius of 16 kilometres of the city centre (Schmitt 1969). Within the next twenty years metropolitan parks systems were developed for New York (much of it well beyond the city boundary), Kansas City,

Denver, Detroit and Chicago, to name a few. The Minneapolis–St Paul system first proposed by Cleveland in 1885 was finally completed in the 1920s and became one of the finest examples of a metropolitan-scale network of parks in the world. In Chicago, the work of another of the great American landscape architects Jens Jensen, in a sense complimented Burnham's civic beautification. In true Olmstedian tradition, Jensen believed in the importance of early contact with nature (he even argued for 10 to 15 acres of open space around every school to let in 'the ever soothing green') and worked to create a parks system based upon the preservation of stream valleys, forests and prairie in their natural state as both landscape contrasts to and recreational areas for the expanding metropolis (Laurie 1979). The revival during the depression years of the belief in nature and the countryside as an antidote to urban ills also helped to maintain interest in metropolitan systems. It was, for example, as pastoral relief for the beleaguered working classes that Robert Moses developed the New York regional parks system in the 1930s (Newton 1971).

The emphasis on the conservation of natural environment in metropolitan parks planning reflected the growing concern about the character and pace of metropolitan expansion. It therefore echoed the idea of natural and country landscape as a relief from the built environment which produced the great urban parks of the previous century. This was especially true of inter-war Britain where the spectre of uncontrolled urban sprawl generated growing reaction amongst planners. One of the leaders of this reaction was Sir Patrick Abercrombie who brought the idea of park space as an antidote and a check to urban expansion back to the forefront of British urban planning. Although never implemented in its entirety, Abercrombie's *County of London Plan* of 1943 proposed an 'Open Spaces and Parks System' which took the metropolitan parks concept to a new level of comprehensiveness, with linear parks, public allotment gardens, parkways and an encircling green belt (Abercrombie 1945).

Abercrombie's ideas had a profound impact on both sides of the Atlantic. In particular they helped to establish the convention of incorporating parks and other open spaces into a broad land-use plan. Although it rarely achieves the comprehensive vision of Abercrombie, this approach has now become the staple fare of metropolitan planning. At its least imaginative it is no more than the use of open space to break up the uniformity of the built environment, and, in the case of green belts, the imposition of boundaries to suburban expansion. Furthermore the shift of park development into the urban fringe reflects the general twentieth-century trend of enjoying natural and rural amenities in, or at least close to, the countryside proper rather than in the contrived landscape and congested atmosphere of urban parks. In this sense, metropolitan park systems are more about taking people into the country than about bringing the country into the city.

However, while metropolitan sprawl and increased personal mobility has enabled millions to satisfy their escapist needs outside the city, the naturalistic urban open space ideal has not entirely evaporated. Indeed one of the main vehicles for its continued expression is the central element of metropolitan expansion itself – suburbia. It was in early suburbia that much of the Olmstedian ideal survived, although, as we shall see later in this chapter, in a derivative form which extended well beyond parks to the very philosophy of suburbia itself. And it is in modern suburbia that any vestige of new naturalistic park development tends to be found. Yet the ideal also survived in the city proper, firstly in the physical legacy of the great nineteenth-century parks themselves and, secondly in the persistent popularity of the country-like settings which they offer to the urban resident. New York's Central Park is a case in point. Despite the various attacks on its original appearance over the years it still provides that feeling of visual and physical escape from the gridiron built-environment of Manhattan which motivated its construction in the first place. And the value of this to New Yorkers is evident not only in its continued use for informal recreation, but also in the recurrent campaigns to preserve and restore its original character (Schmitt 1969).

SUBURBAN ARCADIA

Throughout history suburbs have evinced ambivalent and often strongly contrasting reactions. The residents of ancient Athens, while valuing the countryside for temporary relief from the crowded city, would not have countenanced taking up permanent residence outside it. From early Medieval times until well into the sixteenth century, suburbs were generally the depository of the poor and of the noxious industries that employed them, and were therefore regarded as literally 'sub-urban' and quite unfit for civilised society. And, as suburbanisation has become the principal mechanism for urban expansion and the housing of the middle-class masses in the twentieth century, there has been a recurrent intellectual criticism of and urbane disdain for suburbia's sprawling spatial form, social uniformity and placeless landscape.

Against these negative connotations, however, must be set the persistence of an idealised view of suburbs. Pervading the history of both their development and their image there is a recurrent Arcadian theme. It is a theme which has its classical antecedents in the ring of suburban villas which were built around Augustan Rome by well-to-do merchants and civil servants seeking a bucolic residential escape within daily commuting distance of the city. As a modern ideal extending back over two hundred years, it reveals itself as a bourgeois search for residential separation from the industrial city, which is inspired not by a desire for country living itself but rather by the redefinition of the urban residential landscape through the images and symbols of nature and rurality. In the garden suburb and its derivatives – the suburbia, that is, of the detached, single-family house set in its own private garden amidst leafy surroundings in an exclusively residential community imbued with quasi-village atmosphere – we can recognise the symbolic confusion of city and country. It is a confusion which is revealed in the actual landscapes of modern suburbia, and hence, given its pervasive suburbanisation, in the modern metropolis in general. But it is also a confusion which reveals the depth and persistence of countryside myths in urban society.

The origins of the suburban ideal reside, predictably, in the same social and aesthetic reactions against the industrial city which produced the great urban parks in the early nineteenth century. The parks movement, however, was principally concerned with providing temporary pastoral relief for the masses. The Arcadian suburbs, in contrast, were the creation of a new middle class which sought to escape these masses. Bourgeois domesticity (see Chapter 1) and suburban living went hand in hand. It was a short step to the adaptation of the social and architectural ideals of the country house to the design of the suburban residence and its surroundings. A handsome villa with a large landscaped garden set in park-like scenery was not only aesthetically desirable but also echoed the status of the country gentry to which many of the new bourgeoisie ultimately aspired. Improvements in transportation brought about by new turnpike roads increased the feasibility of suburban living, and by the 1750s the early framework for suburban London began to emerge in Hampstead, Hackney, Highgate, Islington and St John's Wood as well as along the roads connecting these places with the city.

The suburban ideal, then, can be seen as originating as 'a collective creation of the bourgeois élite in late eighteenth century London' (Fishman 1987: 9). Yet, as Fishman has pointed out, this prototype of modern suburbanisation was an Anglo-American phenomenon. By the end of the century, prestigious suburbs had begun to appear around North American cities. Although it is hard to believe today, in the 1790s Brooklyn was a desirable suburban address. With its tree-shaded streets it 'sold nature wholesale to real estate developers', and with its pleasant homes just on the outskirts of Manhattan it soon exuded middle-class ambience (Jackson 1985). Around the same time Bostonian businessmen, too, had begun to move their families into villages on the city's fringe in search not only of a more pleasant visual environment but also of a sense of village community (Binford 1985).

Garden suburbs and cities

By the early nineteenth century the fashion for suburban living had become firmly established amongst the affluent middle classes. The desire for a domestic refuge from the city quickly became a search for status and style. Armed with the romantic aesthetics of the picturesque, architects wasted no time in establishing the design conventions for exclusive suburban villas. Among the most influential in this was John Nash. In 1828 Nash added Park Village to the north-east corner of Regent's Park. Designed in picturesque style inspired by Nash's own work in cottage architecture in his model village of Blaise Hamlet in Gloucestershire, Park Village had an enormous influence on the design of the many exclusive suburbs which sprang up around British cities in the early decades of the nineteenth century (Edwards 1981). Nash's ideas also had a direct impact on American suburban development.

By the 1840s the virtues of suburban living for the discerning American middle classes were being widely promoted by architects, intellectuals, literary figures and social reformers. Nathaniel Hawthorne, returning to his homeland after a stay in England in the 1820s wrote glowingly of the villa parks in which he had lived in the suburbs of Liverpool and London; of Rock Park in Birkenhead with its 'noiseless streets' and 'pretty residences' and of the 'semi-rural scene' of Blackheath Park south of London (Jackson 1985). The social benefits of suburban living were extolled by Catherine Beecher, who rose to national prominence with the publication in 1841 of her best-selling *Treatise on Domestic Economy, For the Use of Young Ladies at Home and at School*, in which she elevated the values of family life and the feminised home to new moral heights. She professed to believe in the superiority of women, but, paradoxically, argued that true womanhood could be attained only through a life devoted to the care of home, husband and children; a family ideal which would thrive best in country or suburban surroundings (Jackson 1985).

Beecher saw a direct link between domesticity and the architecture and landscape of the home. This belief was carried further by the architectural ideas of Andrew Jackson Downing and Calvert Vaux. Downing, as we have already seen, was directly influenced by the English picturesque while, as an English architect Vaux brought this tradition with him to the USA. Downing's essays on domestic architecture helped to establish the detached, single-family house as the American residential ideal. Both Downing and Vaux combined a belief in good taste in house design with an anti-urban preoccupation with natural landscapes, which they applied to their criticism of the gridiron street plan. As an alternative, Downing proposed an ideal suburb in which he pioneered the application of naturalistic landscaping to residential development. 'If a suburb was to be a genuine escape,' he

155

wrote, 'then it must take on aspects of the country' (in Schuyler 1986: 155).

The first successful attempt to build this residential ideal was Llewellyn Park, which, as we saw in Chapter 3, was laid out in 1851 in the Orange Mountains of New Jersey, 18 kilometres west of New York city. Despite its exurban rather than suburban location, the project's Arcadian informality was widely praised by the architectural profession, and was quickly followed by a flurry of development of picturesque suburban communities or 'romantic suburbs' to use Jackson's term (Jackson 1985). The central architectural figures in this were again Vaux and Olmsted. To Vaux's attachment to the ideal of the country residence, Olmsted added his broadening vision of the city. In properly planned suburbs he saw the opportunity to extend his naturalistic softening of the artificial urban environment beyond

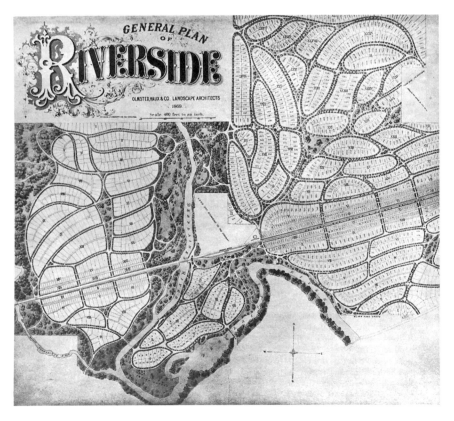

Plate 5.4   Olmsted and Vaux's plan for the garden suburb of Riverside, Chicago, 1869

the public park to the everyday residential landscape. These ideas were first implemented in Riverside, arguably the most influential of all American garden suburbs. On 700 hectares, 15 kilometres by rail from Chicago, Olmsted and Vaux laid out a fully self-contained suburban community. Their central aim, as we can see in the original plan shown in Plate 5.4, was to achieve a sylvan, village-like atmosphere through the use of a curvilinear street pattern, large lots, houses set well back from the street, liberal tree-planting and an extensive parks system (Newton 1971). Now encircled by modern suburban sprawl, Riverside, with the benefit of over a century of tree growth, still retains much of its garden suburb character.

It was, of course, the railway which made the garden suburb a practical proposition, for it permitted development on cheap land and in rural settings, while retaining a daily link with the city. Garden suburbs therefore soon became railroad suburbs, exclusive end-of-the-line communities, in which the ideas first implemented in Llewellyn Park and Riverside were enthusiastically adopted by private developers. Olmsted, Vaux and Company were commissioned for numerous private schemes between 1870 and 1890, including Brookline and Chestnut Hill in Massachusetts, Sudbrook and Roland Park in Maryland, and Yonkers and Tarrytown Heights in New York. In some respects these were exurban communities, hooked into the social fabric of a gentrified country lifestyle. And, as cities expanded, the demand for exclusive residential communities, located in attractive countryside as far from the urban boundary as the rail service would permit, continued to grow. Many of these communities vigorously resisted attempts at urban annexation, preferring instead to preserve their political independence and their village character, an objective which was often reinforced by deed restrictions on unsuitable development (Jackson 1985; Binford 1985).

There is, then, a good deal of semantic confusion over the use of the term 'suburb' to describe communities whose very attraction lay in their spatial separation from the city. Yet their attraction lay also in their proximity to the city and the daily contact that this allowed. The American garden suburb was therefore a conscious attempt to achieve a compromise between these apparently conflicting demands, by creating the illusion of rurality in both location and internal design. As cities pushed further out, the locational illusion became increasingly difficult to achieve and many garden suburbs like Brookline and Roland Park soon became embedded in the urban framework. But the design concepts, indeed the general residential ideal of the garden suburb had become and would remain part of the professional convention of American suburban planning. One reason for this was the changing character of suburban development itself. The introduction of the electric tram in the 1880s opened up large areas of land to speculative tract development which produced not only monotonous residential estates but also attracted the working classes into the suburbs in unprecedented numbers. It was the professional reaction against this, led by Olmsted, together with the growing recognition amongst private developers of the marketability of exclusive middle-class suburbs which ensured the perpetuation of the ideals of the garden suburb in American urban development.

Improvements in transportation, of course, played just as important a role in the building of garden-style suburbs in Victorian Britain. The introduction of the horse-drawn omnibus and tram, the astonishingly rapid spread of the suburban railway network and, at least in London, the building of the Underground, facilitated a rash of speculative suburban development. Most of this took the form of 'by-law' housing; row after row of monotonous terrace housing built through the piece-meal enactment of local by-laws providing development rights to builders. Yet, as in the USA, this in itself increased the middle-class demand for more pleasant and exclusive residential surroundings. 'The richly endowed suburb ... a kind of green jungle of unexpected crannies and enticing green vistas' (Richards 1973: 23); a heavily landscaped 'Victorian pseudo-Arcadia' (Edwards 1981: 25) of large houses set in large gardens became the status symbol of most nineteenth-century cities.

The most notable version of this kind of suburb was Bedford Park which was laid out in 1867 in the London suburb of Chiswick (Plate 5.5). Initially conceived by the developer Jonathan T. Carr, Bedford Park was designed by Norman Shaw (who was the principal exponent of the Queen Anne style in the domestic architecture of the time), who

> arranged around a village green a significant secular-looking church, an arty pub, a row of generously-glazed shops and the tree-lined entrance to radiating avenues of red-brick, Dutch-gabled houses, self-consciously evoking the atmosphere of some mythical early seventeenth century village.
>
> (Taylor 1973: 60).

Bedford Park, however, differed from other Victorian planned suburbs in that it was conceived as an arts and crafts colony (both Carr and Shaw were followers of William Morris). In this sense it was part of the growing movement in favour of radical and reformist planning solutions to the problem of the Victorian city.

At the heart of this was the concept of the garden city, which, while it originated very much as a reaction against suburbia, came to have a profound influence not only on the subsequent development of garden suburbs but also of modern suburbia in general. The idea of the garden city is universally associated with the work of one man: Sir Ebenezer Howard. A relatively ordinary Englishman who worked as a parliamentary recorder, Howard became interested in questions of land reform and ideas about utopian communities. His proposal for garden cities was first publicised

Plate 5.5 Norman Shaw's original sketch for houses in Bedford Park, London

in 1898 in a pamphlet entitled, *Tomorrow, A Peaceful Path to Real Reform* (retitled, *Garden Cities of Tomorrow* in its second edition four years later). In this he outlined a concept which was to have a profound effect on the subsequent development of town and country planning throughout the western world. Howard argued for the need to relieve the pressure of over-crowded cities, a problem which he believed to be of general concern. 'It is well-nigh universally agreed by men of all parties,' he wrote, 'that it is deeply deplored that the people should continue to stream into the already over-crowded cities, and should thus further deplete the country districts' (Howard 1898: 42). He likened the situation to two magnets – the town and the country – each with its own disadvantages; the congested town and the impoverished country. The solution was the creation of a third magnet; a combination of town and country. In the 'town–country' magnet 'all the advantages of the most energetic and active town life, with all the beauty and delight of the country, may be secured in perfect combination' (46) (see Plate 5.6).

Although Howard did propose a schematic layout for his model community, he was more concerned with the principles of the garden city – the general framework of land use and ownership, and the administrative and economic structure – than with its specific morphology. It is important to recognise that Howard's ideas were not conceived in an intellectual vacuum. He did, after all, live in an age of renewed doubts about industrialisation and its urban consequences, and of a rising tide of nostalgia for rural England. Model industrial villages and garden communities had already become a feature of the English landscape. Among Howard's earliest protagonists were George Cadbury and William Lever, builders of the model villages of Bournville and Port Sunlight (Darley 1975). He was influenced also by the land reform ideas of Henry George and the co-operative philosophy of Kropotkin (Fishman 1977). But Howard's main inspiration was the American Edward Bellamy. He became actively involved in the promotion of Bellamy's book, *Looking Backward* (1888) in Britain and used it as a springboard for gaining support for his own ideas about the ideal community. The book had a considerable following in English radical circles and it was within these that Howard tested his early ideas for the garden city.

It was Bellamy's vision of a perfect community, his optimistic notion of freeing workers from manual labour and the general elimination of the misery of the industrial city which attracted Howard. But he came to reject Bellamy's authoritarian socialism and metropolitan centralisation in favour of the small-scale, co-operative communities envisaged by Kropotkin (Fishman 1977). The garden city was therefore conceived as a kind of 'reconstituted village . . . a world of country vivified by the thought and business of town-bred folk' (Creese 1966). Like Kropotkin, Howard believed in self-contained and decentralised communities. Although it was

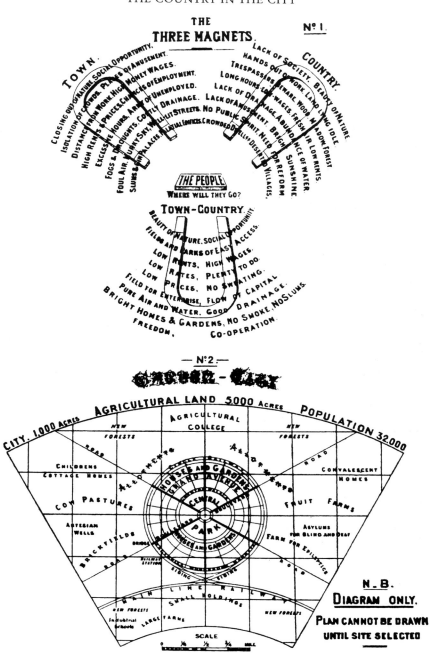

Plate 5.6 Sir Ebenezer Howard's 'Three Magnets' and 'Garden City'

not anti-urban it was an essentially non-metropolitan recipe, with a subtle rural flavour. The garden city was to be small-town in scale and embedded in the countryside. It would, he argued, cause 'the spontaneous movement of the people from the crowded cities to the bosom of our kindly mother earth, at once the source of life, of happiness, of wealth and of power' (Howard 1898: 46). Thus, although spatially separate from the town, agriculture was to be functionally integrated into the garden city economy. Howard believed also that social and economic reform was dependent on the quality of the living environment, and that the beauty of the country should be the guarantee of this quality. 'In 'Town–country' equal, nay better, opportunities of social intercourse may be enjoyed than are enjoyed in any crowded city, while yet the beauties of nature may encompass and enfold each dweller therein' (48–9).

Howard's ideas rapidly acquired influential supporters and in 1899, with the support of the Land Nationalisation Society, he formed the Garden City Association. By 1903 the Association had raised enough money to embark on the construction of the first garden city at Letchworth in Hertfordshire. Although Howard was intimately involved in the planning of Letchworth, it was the ideas of its architects, Barry Parker and Raymond Unwin, which increasingly influenced not only its character but also the subsequent evolution of the garden city concept. While Howard's basic principles were faithfully followed, Letchworth was also an opportunity for Unwin to put into practice his own William Morris-inspired ideas about social and aesthetic unity. It was Morris's pre-industrial revivalism rather than Bellamy's futuristic utopia which inspired Unwin. Much of the design of Letchworth reflected therefore the continued influence of the arts and crafts movement: the medieval village as the historic standard of the ideal community, the necessity for continuity with the past, architectural simplicity and the indispensibility of beauty (Creese 1966). Like many of his contemporaries Unwin was deeply nostalgic about the old English village (Fishman 1977). The philosophy of the first garden city was thus expressed in a style derived from the village, first in housing built in the traditions of the vernacular cottage architecture of south-east England and then in a low-density layout (the standard in all Parker and Unwin developments was twelve houses to the acre) and irregularity of street pattern which evoked the informality and openness of village morphology (Fishman 1977; Relph 1988). This, together with the co-operative ideology of the community, was no doubt responsible for the fact that in the early years it attracted theosophists, vegetarians and folk-dance enthusiasts (Creese 1966).

Although Letchworth embodied the essence of the self-contained industrial community set in the countryside which Howard had proposed, its physical design was as much about bringing the garden into the city as establishing a city in the garden. Great attention was paid to public greenspace – to landscaping and planting – and to the sense of traditional cottage lifestyle through the inclusion of private gardens for each dwelling. For Parker and Unwin these design principles had more to do with providing an alternative to the landscapes of suburban sprawl and urban squalor than with the social ideals of the garden city. This is not to say that Howard's ideal died after Letchworth. Welwyn Garden City, laid out by Louis Soissons on land assembled by Howard in 1919 not far from Letchworth, continued not only the idea of the planned, self-contained community but also took one step further the principles of spacious landscaping and low density development, so much so that some critics derided Welwyn as romantic and anti-urban (see Osborn and Whittick 1977).

However, even while Letchworth was under construction the garden city ideal began to diffuse into a more general urban planning philosophy. Unwin in particular became increasingly active in advocating its application to the control of the growing problem of haphazard suburban expansion (Osborn and Whittick 1969). In 1906 the partnership was commissioned to design a suburban garden village for City of London housing co-operatives at Ealing (Hall 1988). But their most influential achievement was Hampstead Garden Suburb in north London (Plate 5.7). The brainchild of the social philanthropist Dame Henrietta Barnett, Hampstead was orig-inally conceived as a way of preserving the natural landscape of the Heath from encroaching speculative development and of providing all classes with good housing in pleasant surroundings within easy reach of the city (Creese 1966). Parker and Unwin were appointed as architects for the project in 1907 and over the next four years set about designing a suburb with many of the visual effects used at Letchworth: low densities, houses arranged in different patterns, clusters around small greens, cul-de-sacs and footpaths, gardens and shade-trees and a variety of housing types in the domestic cottage tradition, which to this day evokes the cosiness of the English village. Other arts and crafts-inspired architects designed houses at Hampstead, including the principal exponents of the cottage style, Voysey and Scott. The pre-industrial spirit was completed by the addition of public buildings in the medieval style of the Germanic picturesque by Parker and Unwin and by Sir Edwin Lutyens who was responsible for the Central Square, a teutonic anomaly which today stands in sharp contrast to the Englishness of the residential areas.

The absorption of the garden city ideal into the development of garden suburbs was a process that was repeated across the Atlantic. The formation of the Garden Cities Association of America in 1906 stimulated renewed interest in the planning of ideal communities. Although none of the Associ-ation's early plans came to fruition, it did influence the building of one of the most celebrated of American garden suburbs – Forest Hills Gardens, which was laid out on Long Island between 1912 and 1915. With its Hampstead-inspired medieval village appearance and Germanic-style

Plate 5.7 Hampstead Garden Suburb: sketches for 'medium-sized' houses

Station Square, Forest Hills retains to this day the decidedly Utopian tone which prompted its original construction.

American interest in the planning ideals of the garden city was kept alive by the formation in 1923 of the Regional Planning Association of America (Christenson 1978). Prompted initially by the post-war housing crisis, the Association's work soon became a general condemnation of speculative suburban sprawl. The solution, it was argued, lay in the planned decentralisation of urban expansion based on a garden city principle that would restrict the level of population, profit and congestion in each city (Schaffer 1982). The garden city, however, was seen as only part of a comprehensive scheme of regional planning, the leading exponent of which was Lewis Mumford, a founding member of the RPAA, and arguably the most influential critic of the American city. Mumford had been impressed by Howard's approach to decentralisation, but was especially inspired by Sir Patrick Geddes's bio-sociological ideas about the integration of the city with its natural region (Creese 1966). For Mumford (1951), this translated into a means to avoid what he called the 'wilderness of suburbia' and replace it with 'balanced communities, cut to a human scale, in balanced regions, which would be part of an ever widening national, continental, and global whole, also in balance' (15).

Despite these grandiose ideas, the RPAA's immediate concern was the construction of a single garden city. In 1926 it embarked on the now famous Radburn project on a site in New Jersey 25 kilometres from New York City. From the outset Radburn was destined to be an incomplete version of the garden city. It was not based on community land ownership nor did it contain the elements for a self-contained community. But in its internal design it pioneered new principles of suburban development. Central to the 'Radburn Idea' was the belief current amongst social reformers of the time that the family and the neighbourhood were threatened by modern urban life. For Clarence Stein, Radburn's chief architect, the solution was to return to the values of the village community, to a 'safer, more orderly and convenient, more spacious and peaceful community than the conventional city' (Stein 1950: 72–3). Organised around the elementary school and a central communal open space akin to the village green, Radburn's neighbourhood units were explicitly evocative of the small-town landscape. More significantly, it was planned to deal with the problems of the new automobile age. Walkways and underpasses separated pedestrians and cyclists from vehicles, while an emphasis on nature and greenspace in landscaping completed the sense of idyllic isolation from the turbulent world (Christensen 1978).

As model communities garden suburbs and cities revealed the possibilities of the ideal residential environment; of something better than both urban congestion and unplanned suburban sprawl. The construction of Letchworth and Hampstead had an immediate impact on suburban development at the beginning of this century. Although the fashionable garden suburb, as we have seen, had already become a feature of Victorian cities, Unwin and Parker's new pattern of residential design offered fresh opportunities to developers of suburban estates. These were not complete garden suburbs in the sense that Hampstead was intended to be, but rather estates which set out to capture much of the appearance of garden suburbs. What attracted house-builders and their upper middle-class customers was the pre-industrial domestic architecture, natural landscaping and village atmosphere of Letchworth and Hampstead. This became the established style of development for affluent suburban estates in Edwardian Britain. According to Darley, by 1914 there were over fifty housing projects along these lines completed or under construction in Britain (Darley 1975). There were probably more, for most cities, large and small, can today still boast at the least the remnants of leafy, Edwardian 'garden suburbs', with their large mock-Tudor, mievre (pretty) or neo-Georgian houses set in hedged gardens along curving tree-lined streets.

Across the Atlantic, domestic garden-suburb development at the beginning of the century was still moulded largely by Olmstedian design traditions. With the automobile the possibilities for exclusive residential communities on the suburban frontier became almost limitless and new versions of the romantic suburb appeared on the outskirts of most American cities until well into the 1920s (Stern 1981). The best-known of these are Kansas City's Country Club District, begun in 1907, which is famous for being the first garden suburb planned for the automobile age, and Los Angeles's Palo Verdes and Beverly Hills. Yet, at the same time Hampstead did engender considerable enthusiasm within the architectural profession and the house-building industry. Suburban estates in the Hampstead design tradition were constructed particularly around the more traditional eastern cities as well as in English-speaking Canada. In staunchly British Toronto for example, 'packaged suburbs' like Lawrence Park (1910) and Forest Hill (1920s), explicitly Hampsteadian in the romanticism of their architecture and layout, increasingly supplanted less planned patterns of development (Paterson 1989).

Modern suburbia

By the 1920s garden suburb design had become an established instrument of the housing industry, while the subsequent history of suburban development is replete with projects which purport to be garden suburbs. A few of these represent genuine architectural attempts to re-create the principles of a model community. But, for the most part, they are garden suburbs in name only; commercial developments which merely exploit the garden suburb image in order to sell suburban housing to the affluent and the status-conscious. It is in this type of development that the Arcadian

Plate 5.8   Golders Green as Arcadian retreat: 1910 Metropolitan Railway advertisement

escapism of the Anglo-American suburban ideal has been most self-consciously expressed. This is particularly evident in the history of British suburbia. As early as 1910, for example, the Metropolitan Railway (which was responsible for stimulating much of the suburban growth around Edwardian London) was advertising housing estates in which the obvious intention was to cultivate the illusion of a garden suburb with the mock-Tudor, arts and craft-inspired house, the fenced garden in picturesque theme, the shade-trees and the tree-lined street leading to the station, and the scene of domestic peace and harmony, together with the ultimate in rural nostalgia, the lines from William Cowper (Plate 5.8).

For the speculative builder the commercial opportunities of this style of development were obvious, especially with the rapid suburbanisation that resulted from the huge demand for housing and the transportation improvements that followed the Great War. Tree-lined, grass-verged avenues and cul-de-sacs of detached and semi-detached neo-Georgian, cottage-vernacular and mock-Tudor became the stock design of inter-war British middle-class suburbia (Edwards 1981). Housing estate names, and their advertising evoked rural settings and associations while promoting urban convenience. Perivale Wood at Ealing, for example, was 'London's most Rural Estate' yet Ealing Broadway 'with its shops, etc.' was 'close at hand' (in Oliver et al. 1981: 48). House designs (popularised by Ideal Home magazine) combined interior modernity with exterior nostalgia for old England – 'homes fit for heroes', a wartime promise of Lloyd George became popular as a post-war slogan of the developer:

> He made them think that, despite Passchendale and the Somme, despite the coal strikes, despite the millions of unemployed, despite the dreary round of cooking and housework, they, the middle-classes of Britain, were the heirs of Merrie England – a golden age of thatched cottages with roses round the door and Drake playing bowls on Plymouth Hoe.
>
> (Edwards 1981: 129).

A debased version of the garden suburb to be sure and, in reality, a pseudo-Arcadian residential landscape of monotonous predicability. Yet, as a means of housing the general population, even Raymond Unwin regarded this kind of suburban development as not only healthier and more visually appealing than by-law housing, but more economical to build (Unwin 1981). The housing industry was therefore not alone in its attempts to extend the tone of the garden suburb to that of suburbia at large. Much of the responsibility for this rests also with the early urban planning movement of which Unwin was, of course, one of the leading figures.

That the ideas of the garden city/suburb movement became the basis of suburban planning conventions is hardly surprising. After all these were the first significant ideas about the design and planning of the modern metropolis. Largely through the dominance of men like Olmsted and Unwin, they quickly became the conventional wisdoms of the infant professions of landscape architecture and planning at the beginning of this century, a status which they have largely retained ever since. Unwin played a central role in this development. He had a significant influence on British housing policy immediately following the First World War. Through the Tudor–Walters Report on Housing of 1918 and the Housing Manual the following year, the architectural and landscape principles of Letchworth and Hampstead were adapted to the design of municipal (public) housing estates (Hall 1988). From this emerged the low density, cottage-vernacular style of council housing which was to endure until the 1950s and which produced some of the finest working-class housing ever built. The adaptation of Unwin's ideas to the American garden cities movement added the principles of neighbourhood planning and vehicle–pedestrian separation which were pioneered at Radburn.

By the 1950s the ideas of garden city, new town and garden suburb had merged, in varying levels of dilution, into planning standards for suburban residential development which were remarkably similar on both sides of the Atlantic. Low density detached and semi-detached housing, private gardens and public open spaces, curvilinear road plans, cul-de-sacs and crescents, pavements and tree-lined grass-verges (sidewalks and boulevards); these are the familiar images of modern suburbia. In part they represent the planning and building code needs of the fully-serviced, automobile-dependent residential estate, and in this sense are far removed from the Arcadian ideals of the garden suburb. Indeed, as both Edwards writing on British suburbs and Jackson on the USA have pointed out, they have produced a depressing uniformity in the character of many modern suburbs (Edwards 1981; Jackson 1985) (See Plate 5.9). There can be no better illustration of the reduction of the garden suburb ideal to its simplest, cheapest and most easily reproduced form than Levittown, the first mass-produced suburb. Laid out on Long Island 40 kilometres east of Manhattan, Levittown captured the image of the garden suburb with detached homes, built in three basic designs ('Cape Cod', 'Rancher', and 'Colonial'), curvilinear and tree-planted streets (called 'roads' and 'lanes'), landscaped front-yards and fenced back-yards, and public parks (Jackson 1985). This became the standard pattern of North American post-war suburban development.

This pattern has been repeatedly condemned for its social, economic and landscape failings. By the 1960s sociologists were criticising this supposedly Arcadian retreat from the inner city for its lack of community and its thin veneer of middle-class respectability (e.g. Whyte 1956). More recently feminist critiques have pointed to the problems of a patriarchal suburban domestic ideal which isolates women from the mainstream of economic life (Hayden 1984). Commentators on urban design have censured modern

Plate 5.9   Mock Tudor in modern suburbia. It could be any suburb where it snows, but this happens to be Scarborough, Ontario

suburbia for its 'placelessness' and its environmental failings (Jacobs 1961; Relph 1976). Yet the suburban ideal remains deeply embedded in the middle-class psyche. Most people in Britain and North America live in suburbia. A growing percentage of these form a suburban working class associated with the increasing diversity of many older suburbs. Yet the dream of a detached house set in its fenced garden, on a safe, peaceful, tree-lined cul-de-sac still remains strong as middle-class suburbia pushes relentlessly out into the countryside. And, as the monotonous deceit of conventional suburban design becomes increasingly apparent, the design principles of the garden suburb keep reappearing in developments aimed at re-capturing the rural atmosphere of the original suburban ideal. Nowhere is this now more evident than in the current fad of 'neo-traditional' planning which sees a return to the traditional morphology and design of the village and the market town as the solution to the problems of suburban sprawl (Duany 1991).

THE GREENING OF THE CITY

Urban parks are still popular claim Whitaker and Browne (1971), citing not only the current movement to reclaim Central Park for the Olmstedian purists but also the continued attraction of the pastoral character of Regent's and Hyde Parks as well as smaller London retreats like Holland

Park, with its village-green atmosphere completed by Sunday afternoon cricketers. They point also to a renewal of professional enthusiasm for the pastoral and the picturesque in new park design. Of the continued popularity of the urban park as a retreat into at least a version of nature there can be no doubt. And, throughout the shifts in the philosophy and politics of urban parks, the Olmstedian ideal has endured.

Alongside this, however, there has grown up in recent years a new and more comprehensive vision of the role of nature in cities. It is a vision which looks well beyond the issue of public parks to a philosophy which sees nature, not just as landscape relief but as an essential and integral element of urban life. This is an integration of architecture and nature first explored by Sir Patrick Geddes and Frank Lloyd Wright, but brought to the fore in the new ecological consciousness of the 1960s by the exponents of the environmental planning movement, notably Garrett Eckbo, Grady Clay and Ian McHarg. It was Eckbo (1969) who argued for 'total landscape design . . . symphonic relations between structures, topography and trees' as the basis for re-thinking the romantic (and Olmstedian) idea of forcing nature on to architecture. For McHarg (1969) the task was to adapt urban design to natural form and landscape, rather than forcing it into a public park: 'clearly the problem of man and nature is not one of . . . ameliorating the conditions of the grim city . . . it is the necessity of sustaining nature as a source of life' (19). This transcendentalist echo was, to use Grady Clay's phrase, a plea for the 'livable city' (Clay 1973); a new version of the Olmstedian vision in which the humanising effect of nature is achieved not through the aesthetics of the gardenesque, but by the integration of what McHarg called 'natural-process lands' into the urban framework.

This represented a reaction within landscape architecture not only against the uniform and technologically dependent horticultural approach to nature of the established design tradition but also against the use of nature and open space as master planning (Hough 1984). Not that the established traditions evaporated. On the contrary, they still form the basis of much urban design. But the ecological approach did acquire a professional following, and nature-in-cities became in the early seventies somewhat of a bandwagon. There was a flurry of books and conferences on the subject which promulgated an increasingly nature-oriented interpretation of urban design. A key event was the *Nature in the City* Symposium held in Manchester in 1974. Subtitled 'The Natural Environment in the Design and Development of Urban Green Space' it marked the growing emphasis on the importance of putting urban people into daily contact with the natural environment. 'All too often,' wrote Ian Laurie in his introduction to the Symposium's proceedings, 'the developed landscape is biologically sterile and aesthetically depressing . . . Not only is contact with nature too frequently lost within the town, but it is continually being eroded away within the urban fringe and so urban man finds it increasingly further and further out of his reach' (Laurie 1979). For the contributors to the symposium the solution was a comprehensive programme of urban nature and wildlife conservation and renewal, not just in parks and designated areas, but in the ordinary spaces of urban neighbourhoods.

Not surprisingly this approach was, and still is regarded with a good deal of scepticism and no little hostility by urban designers and politicians alike. It was an approach, however, which was entirely in tune with the populist environmental conservation and urban political movements of the times. Conservation groups which had previously paid little attention to the city began to take an interest in urban nature and wildlife conservation. Advocates of community and neighbourhood self-determination seized upon the same idea as a fresh approach to urban renewal. Out of this was born a broadly based movement controlled by a new breed of professional and lay advocates for the 'urban green'. That this was a genuine movement is evident from the rhetoric of the 1984 *Green Towns and Cities Congress, UK/USA*, which was convened by a variety of conservation organisations (including the National Association for Olmsted Parks) and professional bodies.

> In many localities new things are happening, a resurgence of interest in the urban green. It shows up in schemes for the comprehensive renovation of historic parks; in the creation of vest-pocket parks and of neighbourhood commons; in city gardens and nature reserves; in bridleways and river valley parkways; in roof gardens and window-boxes; in new partnerships between the public, voluntary and private-sectors.'
>
> (Dower 1984: 12)

One stream of today's vision of urban green, which is particularly strong in the USA is the re-invocation of the ideals of the Olmstedian park. The Frederick Law Olmsted Society and the National Association for Olmsted Parks have led the campaign to restore the historic naturalistic parks of American cities to their former glory. One example of their influence is New York City's commitment to renovate Prospect Park in Brooklyn (Dower 1984). However, despite the symbolic influence of Olmsted, the mainstream of modern urban green ideology actually tends to reject the formally landscaped park as the vehicle for the greening of the city, emphasising instead a more liberal concept of parks as well as the integration of nature and green space into the everyday urban fabric. To some extent this reflects the continuing influence of McHarg's urban landscape ecology. Yet for current protagonists of urban green, design is a secondary concern. The principal goal is restoring contact between people and nature as the basis for a 'livable' urban environment. Hence the argument for moving nature out of the escapist context of the conventional park, and into urban neighbourhoods.

Nature conservation, therefore, has become a major thrust of the urban green movement. At the professional level it is revealed in a growing body of literature not only on the philosophy but also the methodology of incorporating nature into the urban landscape (Emery 1986; Hough 1984; Laurie 1979; Nicholson-Lord 1987). And the launching in 1988 of a new journal called *Urban Wildlife* affirmed that it was an approach which had acquired a measure of respectability. Much of this literature is British and it is in Britain that urban nature conservation has gathered one of the strongest followings. Since the late seventies a network of nature and wildlife organisations and related government agencies have taken what has been traditionally a countryside issue across the urban boundary. Through the official support of the Nature Conservancy Council, the establishment of several Urban Wildlife Groups and the involvement of organisations such as the Royal Society for the Protection of Birds, the Woodland Trust and Friends of the Earth, urban nature conservation has received growing attention. The County Trusts for Nature Conservation have acquired sites for nature and wildlife reserves, while local authorities (which, with the Nature Conservancy Council have a statutory responsibility for urban nature conservation) have used their powers, albeit on a modest scale, to create officially designated Local Nature Reserves (Harrison *et al.* 1987).

A few authorities have taken a serious interest in this. Until its demise in 1986, the Greater London Council employed a team of five specialists including a senior ecologist to implement ecological planning guidelines. In addition to creating an inventory of the remnants of natural habitats and of the species of wildlife in the greater London area, they have been instrumental in identifying 2,000 sites for nature and wildlife conservation projects. These range from nature reserves on railway embankments and abandoned gasworks to the creation of a 'citywild' nature park on a derelict coalyard at Camley Street behind King's Cross Station (Plate 5.10). Camley Street is a 'small patch of wilderness... vital to the existence of native species of plants and animals in the city' (Countryside Commission 1985c). It is also a focal point for a programme of improving the appearance of London's old canals so as to make them attractive for recreation.

Reclamation and restoration are, in fact, the unifying themes of the urban green movement as a whole; restoration and reclamation, however, not just of the landscape but of people's relationship with nature and of the quality of urban life. Thus this modern version of the old Arcadian idea of nature and green space as the cure for urban ills, is not so much about the appearance of cities, although this is still important, but about the regeneration of urban community. It is this perspective which has broadened the concept of the urban green to include city farms, community garden plots and local food production. A good example of this is the Bronx Frontier Development Corporation in New York, which was

Plate 5.10 Reclaiming nature from urban dereliction: Camley Street 'Citywild' Nature Park, London

founded in 1976 by policeman-turned-ecologist Jack Flanagan amongst the dereliction of the South Bronx. The 'Frontier' is a grass-roots group which provides support to over fifty community gardens which form the basis of a programme for the 'productive greening' of the South Bronx. Some community gardens have become commercially successful (one sells herbs to downtown restaurants), while the Frontier's first project, a municipal composting operation, now sells composted manure from the Bronx Zoo through the best stores in town under the trade name of 'ZooDoo' (Flanagan 1986).

Underlying much of the urban green movement is a strong element of green politics, with its emphasis on the interdependency of local self-determination, personal contact with nature and environmental sustainability. Viewed from this perspective urban greening suggests a radical restructuring of the politics of urban planning, which has little to do with bringing the country into the city. Yet behind the exhortive rhetoric lies an approach to urban rehabilitation which reflects the same fundamental ambivalence about the industrial city and the same faith in nature as its saviour which prompted the first public parks. When today's urban reformers speak of greening as the 'humanising' of the city they echo the early Victorian notion that nature and open space would make for 'a more healthy and moral people' and a more civilised city. The means of achieving this have changed over the years, but the basic principle remains much the same.

6

THE COUNTRYSIDE
MOVEMENT

Underlying the idealisation of the countryside is a strong advocatory
theme. Over the last century or so there has emerged what can simply be
described as a countryside movement; a growing public campaign for the
preservation and conservation of the very images and amenities which have
made the countryside such an attraction. It is not a single, unified movement but rather a diverse collection of individuals and organisations that
have pursued the protection of the countryside from a variety of motives
and perspectives. Add to this the obvious differences in British and North
American approaches and it becomes clear that we are speaking of a
movement in only the most general sense of the word. Nevertheless, the
protection of various aspects of the countryside has become, on both sides
of the Atlantic, a public issue which both mirrors and reinforces the
countryside ideal.

Although they have some features in common, the differences between
British and North American countryside movements are significant. In its
longevity, scope, level of organisation, public support and influence, the
British movement is unrivalled. In no other country is there such an
institutionalised and persistent campaign for the protection of the countryside. But what distinguishes the British movement most of all from its
North American counterpart is the degree to which it has become a
national campaign which focuses on the countryside as a whole as well as
on its constituent parts. For reasons which will be discussed later in the
chapter, countryside conservation in North America has been dominated
until very recently by specific concerns, such as wilderness protection and
historic preservation, and has been more of a local and regional than a
national issue. In this chapter I shall therefore discuss the British and
North American movements separately.

THE BRITISH MOVEMENT

The campaign for the preservation of the British countryside has its origins
in the late Victorian era. This is hardly surprising, given the growing

anxieties of the times about urban industrialism and the concomitant growth of interest in nature and nostalgia for a pre-industrial age. It was a logical step to extend the intellectual, literary and social expression of these values into a political and educational campaign to protect the very features of British life and landscape which they symbolised. It was a campaign which encompassed a number of diverse interests, but which essentially reflected three main concerns: the protection of nature, the enjoyment of fresh air, open space and scenery, and the preservation of national heritage. To a great extent these still define the scope of the countryside movement today.

The protection of nature

Much of the early impetus for countryside preservation was provided by the nature movement. There was already evidence of popular, even working-class interest in the appreciation of nature in early Victorian years. According to Hill (1980) by mid-century, 'in almost every town and village of Lancashire and Yorkshire botanical societies were established to study the area's natural history' (28). By the 1880s several hundred local field clubs with a combined membership of about 100,000 had been established across the country to promote the study of natural history and archaeology (Lowe and Godyer 1983). But it was the formation of national organisations which brought the issue of nature conservation into the political arena. Foremost amongst these were the Selborne Society for the Protection of Birds Plants and Pleasant Places, the British Empire Naturalists Association, the Society for the Preservation of Birds, the Society for the Preservation of the Wild Fauna of the Empire (now the Flora and Fauna Preservation Society) and the National Trust for Places of Historic Interest and Natural Beauty.

These organisations initially arose out of a growing concern about the senseless cruelty and waste of game shooting and the threat of extinction to local plant and bird species resulting from the over-zealous collection of specimens by natural history enthusiasts (Sheail 1976). The Selborne Society, which drew its name and inspiration from the great eighteenth-century naturalist, Gilbert White, promoted the recording and preservation of rare plant species. The Society for the Preservation of Birds (RSPB) was organised initially around a campaign to stop the trade in exotic feathers for women's hats. In its first year, in fact, it was an entirely female organisation. One of the first men to be allowed membership was the writer and amateur naturalist W. H. Hudson, who was instrumental in broadening its objectives to include not only the protection of birds but also that of wildlife in general (Nicholson 1987). The National Trust as we shall see, covered much broader preservational interests, but had as one of its early objectives the acquisition and protection of natural areas.

From these Victorian beginnings the nature movement has steadily expanded into an extensive network of organisations and agencies which have increasingly come to influence the direction of countryside conservation in Britain. In the early decades of this century the crusade for species protection and the prevention of the wasteful and cruel treatment of wildlife continued to be the main focus of nature organisations. But there was also a growing recognition among naturalists that it was the destruction of natural habitats by urbanisation and other changes in land use which represented the main threat to nature and wildlife (Sheail 1976). In 1912 the Society for the Promotion of Nature Reserves (SPNR) was established with the objective of pressuring the National Trust and individual landowners to set aside areas as permanent reserves for the protection of wildlife and nature. However, it was not until the surprising flurry of interest in post-war reconstruction which accompanied the early years of the war, that the SPNR's proposals for a national system of nature reserves finally attracted government attention (Sheail 1976; Lowe and Godyer 1983). In 1949 the government created the Nature Conservancy (since 1973, the Nature Conservancy Council), with a mandate to establish and maintain National Nature Reserves and Sites of Special Scientific Interest (SSSIs), to advise on nature conservation policy and to carry out research and educational work (Blunden and Curry 1985).

From this point on the nature conservation movement went from strength to strength. With the financial and institutional support of the Nature Conservancy, the SPNR gradually broadened its role, first through its sponsorship of the Council for Nature (established in 1958 to promote public interest in natural history) and then through its re-organisation around the rapidly growing number of county trusts for nature conservation. As a federation of forty-four county trusts with a total membership of over 145,000 and responsibility for more than 1,300 nature reserves (Blunden and Curry 1985) as well as the mandate for promoting local interest in the natural environment, the Society has come to see itself as the 'only voluntary body concerned nationally with all aspects of nature conservation' (Lowe and Godyer 1983: 159). Certainly its broader scope, as well as its status were reflected in its change of name in 1981 to the Royal Society for Nature Conservation (RSNC). And in recent years it has continued to expand its activities into new initiatives, such as its programme for on-farm nature conservation, the development of its junior branch, the WATCH trust (which involves children in various nature and wildlife projects) and the 1985 'Tomorrow is too late' campaign to raise £100,000 for habitat protection (Countryside Commission 1985b).

The RSNC, however, has not been alone in campaigning for nature conservation. Indeed, in sheer size of its popular support, it is easily eclipsed by the RSPB, the membership of which has grown rapidly from 8,000 in 1958 to over 400,000 today, and which has become increasingly involved not only in fostering public interest in the protection of rare bird species, but in the monitoring of threats to natural environment in general. However, while the RSPB and the RSNC are the heavyweights of the nature conservation movement, they have increasingly become part of a widening nature network. Some, like the Fauna and Flora Preservation Society, the Royal Society for the Prevention of Cruelty to Animals, the Botanical Society, and the Men of the Trees (which belies its sexist name by admitting women members) are long-established and well-connected organisations. In recent years they have been joined by a proliferation of new nature trusts and associations, as is evident in the list of specialist organisations affiliated with Wildlife Link, the national liaison body for wildlife protection in the UK. These include the Wildfowl and Wetlands Trust (founded by the well-known naturalist, Peter Scott), the Otter Trust, the Woodland Trust and the People's Trust for Endangered Species (Wildlife Link 1990).

Access and amenity

Much of the early interest in nature conservation stemmed from an attraction to its aesthetic and psychological benefits. It was as much a concern for the protection of the natural beauty of the countryside as it was for the preservation of nature itself. The amenity value of nature has in fact long been a central feature of the nature conservation movement. The rise of late Victorian local field and botanical societies was prompted by the attraction of natural scenery as well as by a serious interest in the study of flora and fauna. Many of these societies saw nature study as an activity which went hand in hand with walking and rambling in the countryside, a pastime which, as we saw in Chapter 4, grew rapidly in popularity during this period. The growing interest in the countryside, and especially in relatively wild areas like moorlands, for recreation, therefore became another focus for the early countryside conservation movement. In this the main issue was the preservation and accessibility of nature and scenery for public enjoyment and appreciation.

The initial impetus for this movement came from a group of upper-class, radical liberals, who in 1866 formed the Commons Preservation Society. Generally regarded as the first countryside conservation organisation, this society was strongly influenced by the land reform ideas of the day (John Stuart Mill was a founder member, while the Liberal MP and land reformer G. J. Shaw-Lefevre was its principal spokesperson). Its main platform was the protection of commons from urban development, agricultural enclosure and hunting reserves (Lowe and Godyer 1983). Among its early achievements was the preservation of Hampstead Heath and Epping Forest (Blunden and Curry 1985). A parallel campaign was also waged by the rambling associations which were springing up across the country. Some, like the Sunday Tramps were led by leading figures of the conservation movement, such as the historian G. M. Trevelyan and the rural novelist William Meredith. But much of the political impetus came from the largely working-class rambling clubs of northern industrial areas, such as the Liverpool Hobnailers, which became organised around a 'people's rambling movement' (Hill 1980: 24) to force the issue of public access to the moorland hunting reserves of the landed gentry. The formation in 1905 of the Federation of Rambling Clubs marked the beginning of a national campaign not only to promote country rambling as a recreational activity, but also to build legislative pressure for rights of access to open countryside (Stephenson 1989).

It was, however, with the huge growth in the popularity of countryside recreation that resulted from the new mobility of the inter-war period, that the movement for the protection of and public access to scenic areas gathered real momentum. Most of the initial impetus for this came from the campaign for access to the northern moorlands and mountains. This was led by the federations of rambling clubs formed in Manchester, Liverpool and Sheffield during the 1920s. In 1927 the Manchester and Sheffield Federations convened a countryside and footpath conference at Hope in Derbyshire, which was also attended by the Liverpool Federation, the Peak District and Northern Counties Footpath Preservation Society, the London Federation of Ramblers, the Commons Society and the Society for the Control of Abuses of Public Advertising. At this 'a general expression of opinion and policy was obtained on the preservation of footpaths, rights of way, etc., litter, nuisance, "uglyfication" of the countryside, access to mountains and moorlands' (in Stephenson 1989: 81). The main concern of the rambling federations was to lobby for legislation guaranteeing access to privately controlled land. There was, after all, a strongly socialist ideology underpinning the access movement, a fact which was symbolised in the mass trespass of Kinder Scout in the Derbyshire Peak District in 1932. Although it was actually organised by the British Workers Sports Federation, a communist organisation with no particular links to the rambling federations, this event served to galvanise the rambling movement. Increasingly frustrated by the failure to obtain legislation on access to mountains, the rambling federations moved towards the formation of a national organisation and in 1935 the Ramblers' Association (RA) was formed.

During the late thirties and early forties the Ramblers' Association played the central role in the growing campaign for legislation to protect and make accessible the recreational amenities, not just of the northern mountains, but of the countryside in general (Hall 1976; Blunden and Curry 1990). Not only did it reflect the growing popularity of open-air recreation, but it also had the direct support of many Labour MPs. It was perfectly poised therefore to influence the legislative process when Labour took power in 1945. The late Tom Stephenson, the leading figure in the

RA since the war, describes in his history of the rambling movement his own role in drafting the terms of reference for the government's Committee on Footpaths and Access to the Countryside (the Hobhouse Committee), while he was press secretary in the Ministry of Town and Country Planning. The final report of the committee, says Stephenson, 'bore the distinct marks of the RA's hobnailed boots all over it' (Stephenson 1989: 202). This led to the passage of the National Parks and Countryside Act 1949 (of which more a little later in the chapter) which went much, although not all of the way towards meeting the objectives for which the movement had fought for so long.

The Ramblers' Association was not alone in this fight. It had the support of the southern-based and more conservative Commons, Open Spaces and Footpaths Preservation Society (the renamed Commons Preservation Society) which, during the thirties, took an increasing interest in the protection of public rights of access not just to commons but also to customary rights of way across private property. It was, for example, responsible for initiating the protracted campaign for access to Manchester Water Authority reservoir property in the Lake District and was actively involved in the deliberations of the various committees which culminated in the 1949 Act (Stephenson 1989). The RA's work was also complemented by that of the burgeoning outdoor recreation organisations, in particular the Youth Hostels Association (YHA). Formed in 1930 to provide inexpensive accommodation for young ramblers and cyclists, within two years the YHA had a membership of 20,000 (Hall 1976). It immediately became embroiled in the access issue, appearing at public meetings alongside the RA and the Commons Society. Indeed by the 1940s it had become the main ally of the RA (Stephenson 1989).

Although the National Parks and Countryside Act established the principle of public access to open country, the statutory limitations on this, especially those relating to the rights of private landowners within national parks, have provided a continuing *raison d'être* for the outdoor recreation movement. Much of its subsequent activity has been directed at safeguarding statutory rights of access threatened by the actions of leading landowners such as the Forestry Commission, the regional water authorities and farmers. The Ramblers' Association still leads the way in this campaign. With over 40,000 members organised into 200 local rambling groups, it has been able to count on strong support for its opposition to specific access issues and its rallies for the improvement of recreational amenities, such as the provision of long-distance footpaths. The RA, however, is part of a growing network of organisations involved in promoting the recreational use of the countryside. The YHA continues to collaborate with the RA in the access campaign as well as promoting an interest in the countryside through its Field Study Hostels. To these have been added a number of new groups, including the Byways and Bridleways Trust and the Long Distance Footpaths Association. Even the Sports Council now takes an interest in promoting countryside walks. The contemporary access movement is completed by the various organisations which represent more specialised sporting activities, such as camping, angling, horse-riding and field sports, although these are often at odds with the interests of the more conservation-minded ramblers.

The preservation of heritage

In addition to originating in a campaign for nature conservation and the protection of recreational amenities, the British countryside movement also developed out of late Victorian nostalgia. Inspired by the traditionalist philosophies of Morris and Ruskin it produced a mentality amongst influ-ential elements of the British social establishment which sought the preser-vation of all that represented an older and supposedly gentler order. This was directed at both town and country; at the preservation of both the built and the natural environment. Much of the initial activity was con-cerned with the preservation of old buildings and historic sites. Morris and Ruskin, in fact, were themselves directly involved in the establishment of the Society for the Protection of Ancient Buildings. Yet, in many ways, it was the countryside which, as we saw in Chapter 2, stood as the symbol of the true England, for it was the countryside that was the depository of so much of the nation's pre-industrial heritage. With its natural scenery and wildlife, its Norman churches and Tudor villages, its stately homes and historic sites, rural England became the logical focus for much of the heritage preservation effort.

From its beginnings in the late nineteenth century until the present day, the heritage preservation component of the countryside movement has been overwhelmingly dominated by the National Trust. Founded in 1895 at the instigation of Robert Hunter, the secretary of the Commons Preser-vation Society, which itself was becoming increasingly frustrated at its inability to secure protection for common land, the Trust was set up to acquire and manage land and buildings for the benefit of the nation. Its incorporation of both built and natural environment under the single banner of national heritage was reflected in the Trust's full title: 'The National Trust for Places of Historic Interest and Natural Beauty'. It quickly attracted the support of the intellectual and artistic establishment, and, in contrast to the ramblers' associations, of the landed gentry, who saw the Trust as both an ally in the maintenance of the traditional rural order and a solution to the increasingly precarious financial position of their estates (Lowe and Godyer 1983; Tunbridge 1981).

Such was the status and influence of its leading members that in 1907 the Trust became a statutory body by Act of Parliament. This gave it the power to declare its properties inalienable and to enact by-laws for their protection and use. It also gave it the effective status of a public agency, even though it continued (as it still does) to be a voluntary organisation (Lowe and Godyer 1983). This bestowed a respectablity on the Trust, which enabled it to attract donations of both money and property. By 1917 it had acquired 2,560 hectares, the lion's share of which comprised country estates in the south and scenically significant pieces of land in the Lake District (Tunbridge 1981). Since then it has expanded its holdings, to the point that it is now the largest single private landowner in England and Wales, with over 1 per cent of the land surface and more than 200 historic buildings. Its rural properties fall into four main categories: stately homes and gardens (most of which come to the Trust as windfall bequests, especially since it has been possible to use this as a means to avoid death duties), historic buildings and sites, scenic areas, and nature reserves. Mainly through its acquisition of large estates, it also owns seventeen villages in their entirety, which it preserves as significant examples of traditional rural settlement and architecture (Tunbridge 1981). One example is Lacock in Wiltshire, which contains many half-timbered houses with wattle and daub infilling dating from late medieval times including a four-teenth-century cottage in cruck-frame construction. During the Second World War, it was selected to illustrate English village life in a publication circulated abroad to show that the true Britain was still thriving (Woodruffe 1982). Such is the symbolic power of heritage preservation!

In recent years the Trust has moved to a more selective process of acquisition, reflecting a desire for properties which are more representative of national heritage. This approach has extended to scenic areas, especially in National Parks and in Areas of Outstanding Natural Beauty where the Trust is a major landowner, and where, in many respects, it acts as the principal agent for ensuring the preservation and public accessibility of natural and scenic landscapes. Indeed, over the years the Trust has been a leading proponent of national parks, a position which has often led to some ambiguity over the relationship between its central mandate of pre-serving areas for their heritage value and promoting their use for recreation (Tunbridge 1981). In post-war years, too, the Trust has been increasingly drawn into broader issues of countryside conservation and thus into more ambitious projects like Enterprise Neptune, the national appeal launched in 1965 to acquire properties along coastlines threatened by holiday devel-opment and erosion (Lowe and Godyer 1983)

The National Trust, however, remains the establishment organisation of the countryside movement, still reflecting the essentially conservative, olde England values of the late Victorian age. Although it is by far the most important of the heritage organisations, it is not alone in this work. At the national scale, it has been joined by a number of specialist heritage organis-ations such as the Victorian Society, the Georgian Group, the Friends of Friendless Churches, and the recently-formed Historic Farm Buildings

Group. In addition, there is long tradition of local preservation groups which have grown up around campaigns to preserve the traditional character of rural areas, villages and buildings. A number of these emerged in the 1930s as local branches of the Council for the Preservation of Rural England. Since the Second World War, however, there has also been a growth of support at the local level for the protection of the aesthetic heritage of rural settlement. Indeed, it is at the local level that preservationist sentiments appear to have become the most intense. Playing a central role in this is the Civic Trust, which was formed in 1957 to provide advice to local groups on the protection and improvement of the built environment and to foster high standards of civic design. Although it is primarily concerned with urban conservation, many of the societies that are registered with it are based in country towns and villages (Civic Trust 1990).

A common purpose

The British countryside movement, then, has evolved from a wide diversity of interests represented by a plethora of organisations, each with their own specific agendas. Yet, what is striking about the movement, and what distinguishes it so clearly from the North American experience, is the extent to which there has developed a communality of interest in the future of the countryside in general. While it is not possible to speak of a unified movement, it is appropriate to describe it as one which has attained a high level of integration and sense of common purpose.

In fact, this began quite early in the movement's history. The first signs of a sense of common interest amongst the various groups that had begun to campaign for the protection of their particular aspect of the countryside came in 1898. The organisation of a conference by the Society for the Prevention of Abuses in Public Advertising brought together the Commons Preservation Society, the National Trust, the Selborne Society and the Society for the Protection of Ancient Buildings. The objective was to discuss the possibilities of co-operating 'in defense of the Picturesque and Romantic Elements of our National Life' (Sheail 1981). Subsequent meetings convened by other groups attracted a widening circle of organisations, including the rambling associations. These prompted a number of suggestions for a national body to co-ordinate the countryside protection campaign, culminating in architect and town planner Sir Patrick Abercrombie's proposal in 1926 for a National League for the Preservation of Rural England. Abercrombie envisaged a joint committee which could bring together into a single campaign the various bodies concerned with different aspects of countryside preservation (Sheail 1981). In the same year the Council for the Preservation of Rural England (CPRE) was founded, with Abercrombie as its first Honorary Secretary, and with 24 organisations as constituent members and a further 41 affiliated societies. This was followed

Plate 6.1 The village preserved and kept free of tourists' cars: Castle Combe, often claimed to be the most beautiful village in England

in 1927 and 1928 by the formation of parallel organisations in Scotland (the Association for the Preservation of Rural Scotland (CPRS)) and Wales (the Council for the Preservation of Rural Wales (CPRW). For Abercrombie, whose personal and professional concern was for good planning, the issue was not simply the preservation of the countryside but rather the development of a planning system which would permit growth in the countryside 'and yet preserve its beauty either substantially as it is or in a changed form' (Abercrombie 1926: 50).

The new council immediately launched a campaign for the regulation of ribbon development. The leading figure in this was Clough Williams-Ellis who became very much the public voice of the CPRE in its early days, particularly through his book, fittingly entitled *England and the Octopus*, in which he railed against the 'planless scramble' of urban sprawl (Williams-Ellis 1928). The Council also moved quickly to mobilise its member organisations behind the campaign for more general controls on development in the countryside. In October 1929 it convened at Ambleside in the Lake District a National Conference for the Preservation of the Countryside, at which it was resolved to call for the extension of statutory planning to all rural areas. Over the next two years the CPRE acquired growing support in Parliament. Its advocacy of Rural Amenities Bills led to the passage of the Town and Country Planning Act in 1932 and three years later of the Restriction of the Ribbon Development Act (Sheail 1981). These went only some way towards meeting the Council's objectives, but they did represent the official acceptance of the principle of regulatory land use planning as the basis for countryside preservation.

Within a decade of its formation the CPRE had become an effective focus for an otherwise fragmented movement. In addition to leading the way in lobbying for town and country planning legislation, it played a central role in the campaign for national parks which was gathering momentum in the 1930s. Prompted by the Friends of the Lake District, it collaborated with the CPRW in establishing the Standing Committee on National Parks in 1935. Composed as it was of representatives of the Ramblers' Association, the Commons Society, the Youth Hostels Association, the Zoological Society, the SPNR and other related groups, it thus brought together the two streams of nature conservation and recreational amenity to form a common front. From this emerged a *Manifesto for National Parks in Great Britain* which formed the basis for both a parliamentary campaign and the popularisation of the national park idea (Council for National Parks 1986). As we have seen, the Ramblers' Association and other recreational organisations were independently involved in the campaign for national parks. However, it is clear that it was the work of the Standing Committee which kept the issue to the fore, especially during the war years, and which helped to ensure the passage of the National Parks and Countryside Act in 1949 (Blunden and Curry 1990).

In the post-war era, the CPRE (together with its Scottish and Welsh counterparts) has become the main voice of the countryside conservation movement, monitoring national policies and commenting on official proposals which affect the countryside, as well as continuing to co-ordinate the push for further protective legislation. Much of its support comes from its county branches, which have formed the general membership backbone of the organisation from its early years. There are now 47 county branches, which play a key role in monitoring and informing the planning process at the local level. Most have close links to county planning authorities, both through representation of county councils on branch committees and reciprocal branch representation on county planning and countryside committees (Lowe and Godyer 1983). This level of official recognition and influence has been matched at the national scale, for the CPRE has long been represented on all the main government bodies which are involved with countryside policies. This has not, however, diverted the organisation's attention from its original mandate of leading the campaign for the general protection of the countryside. Over the years it has led the fight against the intrusive impact of military training grounds, open-cast mining, motorways, reservoirs, power lines, afforestation in National Parks and development in green belts. It has campaigned for the preservation of hedgerows and the safeguarding of farmland from development. It has launched the Council for Environmental Conservation and played an influential role in the establishment of the Countryside Commission (as well as the campaign to save the Commission from recurrent threats of dismemberment by the Thatcher government).

With a name change in 1969 to the Council for the *Protection* of Rural England, followed by the adoption of 'Campaign for the Countryside' as its operational motto, the CPRE has become more conservationist and propagandist in its activities. In its 1989 annual report, the campaign diary includes opposition to a superstore in rural Berkshire, a public meeting to oppose a theme park at Woburn Abbey, a call for a public inquiry into the rail route for the Channel Tunnel, and the mobilisation of opposition to new villages in the Wessex Downs Area of Outstanding Natural Beauty. Ranging as it does across such diverse issues as agricultural intensification, environmental protection, rural housing, village viability, farm diversification, water resources, energy development, transportation and forestry, the CPRE now appears to embrace every possible aspect of countryside conservation.

The rise of the CPRE is indicative of the increasing level of collaboration over the past half-century or so between the various interests that have defined the scope of the countryside movement. This has been achieved through participation in conferences and joint campaigns, regular contacts between organisations, cross-membership ties, and formal alliances in umbrella groups (Lowe and Godyer 1983). Occasional conferences, which, as we have seen, played a key role in the early consolidation of the countryside movement, have continued to be an important form of collaboration. A good example is the series of Countryside in 1970 conferences, held between 1963 and 1970, in which virtually every organisation with an interest, vested and otherwise, in the future of the countryside participated. It was from this conference that the proposal for the Countryside Act 1968 and the consequent establishment of the Countryside Commission emerged.

It is the links between groups, from exchanges of literature to formal contacts within umbrella organisations, in which the level of collaboration around the general issue of countryside conservation has been most effective. In their survey of British environmental groups, Lowe and Godyer (1983) discovered that, despite the existence of a measure of inter-organisational conflict and rivalry (for example between the ramblers and the field sports enthusiasts), there is a generally high level of co-operation which is described in the chart in Plate 6.2. Much of this occurs through contact between group officials and staff who routinely consult each other and thus work closely together. This is furthered by the cross-membership links between groups, especially where it involves, as it frequently does, leading members of one group holding executive positions in others. Inter-group contact appears to be particularly important at the local level, where individual interest groups frequently act as a network, informally organised to ensure communality of interest in the protection of local amenities.

At both local and the national level, much of the contact between groups occurs through the CPRE, the National Trust, and the RSPB which have become the established points of convergence of interest in the countryside. The CPRE, as we have seen, serves to mobilise and focus support for countryside protection, while the National Trust has increasingly come to act as the patron of a broad range of countryside campaigns and projects. That there has been a growing desire for co-operation is further reflected in the growth of umbrella groups in recent years. The broadest ranging of these is undoubtedly the Council for Environmental Conservation (CoEnCo), which was launched in 1969 by the CPRE to co-ordinate the whole environmental lobby. CoEnCo in turn established Wildlife Link in 1980 to act as the liaison body for all the groups concerned with wildlife protection, including the broader countryside amenity groups such as the Ramblers' Association, the National Trust and the CPRE. It also serves as a link with the Nature Conservancy Council. A parallel body, Countryside Link, consisting principally of recreation and amenity groups, exists to act as a forum for organisations concerned with landscape protection and access, and as a link with the Countryside Commission. The other major umbrella group, and, after the CPRE, the oldest, is the Council for National Parks, which was formed in 1977 out of the Standing Committee on National Parks, and which has since co-ordinated a vigorous campaign

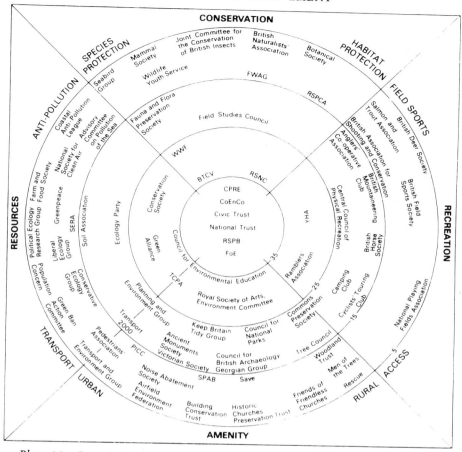

Plate 6.2 Organisation of the British environmental lobby. Groups have been allocated to different bands according to the number of contacts attributed to them (from Lowe and Godyer 1983: 81)

to promote public support for the protection of existing parks as well as the creation of new ones.

Much of the apparent convergence of conservational activity around a general countryside movement is due in large part to the movement's own success in obtaining statutory recognition, first in the form of conservation and planning legislation, but, more significantly, through the establishment of government conservation agencies (otherwise known as 'quangos' or quasi non-governmental organisations). The most influential of these agencies as far as this discussion is concerned is the Countryside Commission. In many respects, it is this commission which has led the way in the maturation of countryside conservation into an integrated movement.

Formed by the Countryside Act 1968, the Countryside Commission for England and Wales (there is a parallel body for Scotland) replaced the National Parks Commission, and thus marked a new phase in the long history of the access and amenity issue. Like its predecessor, it was advisory to the government, but its mandate extended to the countryside as a whole rather than just to National Parks. In 1981, under the Wildlife and Countryside Act, its influence over the countryside movement was significantly increased by its reconstitution as a grant-in-aid body; in other words an agency which had the power to develop its own projects and to provide direct support to voluntary organisations. Over the years, the Commission has carved out its own policy and taken an increasingly independent and pro-conservation position on a number of controversial countryside issues, a situation which has led to recurrent government proposals for its reorganisation and even its total abolition. Such is the public and professional profile of the Commission, however, that it has survived these attacks and steadily expanded its activities.

Today its basic mandate is 'to care for the natural beauty of the country-side of England and Wales and to help people enjoy it'. It does so firstly through the designation of National Parks, Areas of Outstanding Natural Beauty, Heritage Coasts and national trails; secondly through grant aid to local authorities, voluntary bodies, farmers and other landowners to sup-port action on conservation and recreation; thirdly through policy advice to local and national government and other public bodies; fourthly through research and experimental work in conservation and recreation provision; and, finally, through the promotion of public appreciation and understand-ing of the countryside. The record in achieving the objectives implied by these strategies, as one would expect, has been mixed. On the one hand, there can be no doubt that the Commission has had a significant impact on the development of countryside recreation. It has also been successful in increasingly influencing the treatment of countryside landscapes by the local planning process. On the other hand it has been frequently frustrated by its inability sufficiently to influence government policy both at the local and national level. This is reflected for example in the continuing problems over access and land alienation in National Parks, and in the protracted campaign to obtain approval for the designation of new national parks and other areas of attractive and popular countryside. It is reflected, too, in the inherent weaknesses of the Wildlife and Countryside Act, which gave the Commission the responsibility for establishing management agreements for landscape conservation with farmers and other landowners without providing the necessary statutory and financial resources.

THE NORTH AMERICAN EXPERIENCE

Clearly, the campaign for the countryside will reflect the specific values and processes which have led to its idealisation. The North American ideal, as we have seen, has been shaped by many of the same influences as its British counterpart. But, as we have also seen, there are a number of respects in which North American attitudes to countryside have emerged out of very different historical circumstances. The absence in North America of a countryside movement of the level of integration and public profile which has developed in Britain reflects this fact. Firstly, North America has lacked the country-based social and intellectual élitism and associated class struggle which characterised so much of the early British countryside movement. Secondly, North American settlement history has created a landownership structure within which public access to the settled countryside has not been the political issue that it has in Britain. Thirdly, North America has not possessed the same kind of establishment-driven nostalgia for the countryside as a symbol of national identity and order which has influenced much of the British movement. Nor has it exhibited the same sense of threat of disappearance of the countryside in general that has so institutionalised the British countryside campaign. In large part, this reflects the prevailing North American myth of the limitlessness of land, which has long shifted natural and scenic preservational attention to the wilderness frontier and the threat of *its* loss, rather than to the more immediate countryside. But it is also a function of the strength of an agrarian ideology which has valued the countryside as a working landscape. Finally, the sheer diversity, political, cultural and geographical, of the North American continent has not only precluded the possibility of an integrated countryside movement, even at national and regional scales, but also makes it highly unrealistic even to attempt to recognise one.

The preservation of wilderness

The rise of nature worship as one of the defining ideals of American culture in the mid-nineteenth century was more than just an intellectual movement. It was also a foreboding of impending loss; a melancholic anticipation of the advance of civilisation and, with it, the disappearance of the natural world. From these sentiments emerged the call for the preservation of what was left of this world, and especially of what remained of the uniquely American symbol of that world, the wilderness. The first proposals for wilderness preservation have been attributed to the painter George Catlin, who in 1832 suggested a national park to preserve plains Indians and buffalo from the advancing settlement frontier. Over twenty years later, Thoreau was putting forward the idea of 'national preserves' as a means of 'keeping a few places wild' (in Nash 1967). Thoreau's arguments for preserving the north-eastern forests were echoed by others, including the nature-loving Bostonian lawyer Samuel Hammond, who in 1857 proposed reserves of 'wild land' in the Adirondacks. Two years later, the first organised campaign for wilderness preservation came from the Northwoods Walton Club, which called for laws 'protecting our Northern Wilderness . . . as a vast and noble preserve' (Nash 1967: 117).

The campaign for the Adirondacks led eventually to the establishment of the Adirondack State Park in 1894. But it was in the west, in the Rockies and the high Sierras, that wilderness preservation and, more particularly the movement for national parks which it stimulated, came of age. The significance of the western wilderness lay not only in its scenery, but also in the fact that it represented the last real wilderness frontier. Moreover, it was a wilderness of monumental natural grandeur, which was perceived as a symbol of national greatness and identity every bit the equal of the cultural artifacts of the Old World (Runte 1979). It was visited by writers and artists who brought back to the east reports and images which immediately captured the public imagination and stimulated public concern for its protection. The main focus of the movement for the preservation of the western wilderness was the Yosemite Valley in the Sierra Range. Following the discovery of the valley in 1851 and especially of the giant redwoods which occupied it, Yosemite became a magnet for the growing number of travellers who were coming to see the wonders of western scenery for themselves. Initially it was the redwoods alone which stimulated arguments for the valley's protection. At first, British botanists, incredulous of the size of the trees, doubted their existence, but once convinced, threw their support behind the campaign to protect the redwood groves from the activities of private landowners. Concern over the redwoods, however, quickly spread to the scenery of the valley as a whole and thus to the benefits of preserving it for public enjoyment rather than private profit (Runte 1979). Intense lobbying by a group of concerned Californians led to the passage of the Yosemite Act, which granted an area of the valley to the State of California as a park, for 'public use, resort and recreation' (in Nash 1967: 106). Thus the national park idea was born, in fact, if not in name.

The first official National Park was established at Yellowstone in 1872. But the motivation for Yellowstone was the protection of what were regarded as natural curiosities – the hot springs, geysers and canyons – from private ownership, rather than the creation of a wilderness reserve. It was Yosemite, therefore, which continued to capture most of the attention of the preservationists. Among these was Frederick Law Olmsted, straight from his success with New York's Central Park who in 1863 was appointed one of the first commissioners for the Yosemite Valley. In his advisory report to the California Legislature, Olmsted argued that the State had a duty to preserve the natural scenery of Yosemite for public enjoyment and relaxation. However, in a premonition of later controversy over the purpose of national parks, he also promoted the aesthetic appreciation of its natural features, rather than merely treating it as a spectacle (Nash 1967).

The central figure in the campaign for Yosemite, and probably the most influential individual in the development of the national park idea, was John Muir (Fox 1981). Son of a Scottish immigrant homesteader in Wisconsin, Muir took an early interest in nature. Deeply affected by the writings of Emerson and Thoreau, he increasingly based his life on transcendental philosophy. In the late 1860s, after a few years at the University of Wisconsin, followed by a spell as a successful inventor of mechanical devices, Muir decided to explore the wilderness. He hiked widely, usually alone, carrying with him a volume of Emerson's essays. These explorations, especially those which took him along the high trails of the Rockies, confirmed his belief in transcendentalism. For him solitude and simplicity were the prerequisites for appreciating nature in its wild state. Above all, the wilderness was to be appreciated on a spritual level. Muir believed that he had a mission to disseminate the virtues of his view of wilderness appreciation to his fellow citizens. He also began to lecture and write in support of wilderness preservation. His transcendental values were blatantly evident in an early article in the *Sacramento Record-Union* in 1876 entitled, 'God's First Temples: How Shall We Preserve Them?' But his main impact on the national parks movement came in 1890 when at the urging of its editor, Robert Underwood Johnson, he wrote two articles in *Century Magazine* promoting Yosemite as a model for a system of national wilderness parks (Muir 1890).

Muir argued for parks from both ecological and recreational perspectives. Wilderness was to be preserved both for its own sake and for its enjoyment by an appreciative public. With the publicity afforded by a national periodical of the stature of *Century Magazine*, and with other articles in influential magazines, most notably *The Atlantic Magazine*, Muir's proposals quickly attracted official attention. Following a brief period of lobbying in Washington, the Yosemite Park Act, together with similar acts for two much smaller neighbouring parks, Sequoia and General Grant, was passed in 1890 (Runte 1979). However, part of the official motivation for creating national parks was that they contained valuable reserves of forest, minerals and water. And it was the growing tension between the principles of resource conservation on the one hand and wilderness preservation on the other that kept the campaign going. In 1892 Muir founded the Sierra Club with a group of Californian university professors. Dedicated to 'exploring, enjoying and rendering accessible the mountain regions of the Pacific Coast' (Nash 1967: 132), the club became a focus of the campaign to protect National Parks from resource exploitation. Under the club's aegis, Muir continued to promote the appreciation of Yosemite, particularly

Plate 6.3 'God's first temples': Yosemite National Park, California

through his leadership of numerous hiking and camping trips at which his lectures on the natural wonders of the park became famous. But he also led the growing political fight with those who took a more utilitarian view of national parks, in particular Gifford Pinchot, head of the US Forest Service and erstwhile ally of Muir. Pinchot saw the justification for national parks principally in terms of the conservation of economic resources. Muir, on the other hand, argued that the enjoyment of scenic beauty constituted just as legitimate a use of public land (Strong 1988). 'Thousands of tired, nerve-shaken, over-civilized people,' he wrote in *The Atlantic Magazine* in 1898, 'are beginning to find out that going to the mountains is going home; that wildness is a necessity; and that mountain parks and reservations are useful, not only as fountains of timber and irrigated rivers, but as fountains of life' (in Nash 1968: 71).

The dispute between these opposing views came to a head with the infamous Hetch-Hetchy controversy. In 1906 the city of San Francisco renewed an earlier application to secure water supply through the construction of a reservoir in Yosemite's Hetch-Hetchy Valley. Not surprisingly, this prompted an immediate response from the preservationists, the most notable aspect of which was that it brought other groups into the national parks campaign. The Sierra Club obtained the support of the American Civic Association which had already become a rallying point against river-diversions in the east, especially at Niagara Falls. They were joined by another eastern organisation, the Appalachian Mountain Club and in 1906 this east–west alliance of preservationist interests formed the Society for the Preservation of National Parks with Muir as its first president (Jones 1965). Although the Society ultimately failed to prevent the construction of the reservoir, the intense and highly publicised campaign against it significantly increased public support for national parks, and indeed, for scenic preservation in general.

Much of the support for national parks now came from the rapidly growing outdoors and tourist movements. Muir had already recognised the link between scenic preservation and outdoor recreation, while other leading members of the parks movement, notably J. Horace McFarland of the American Civic Association and Allen Chamberlain of the Appalachian Mountain Club argued that the promotion of national parks as a rec-reational amenity would be the most effective way of protecting them (Runte 1979). As improved transportation made the Rockies and other wilderness areas more accessible the scenic wonders of America could be promoted more successfully than ever before. 'See America First' became the nationalist slogan of a burgeoning scenic tourism industry led by the railroad companies. These, too, supported the campaign for national parks. In 1916, following vigorous lobbying led by McFarland and by Stephen Mather of the Sierra Club, the National Parks Service was established, with a clear mandate to manage and create parks in the spirit of the wilderness ideal (Runte 1979).

The inter-war years saw a broadening of the wilderness movement beyond its western roots, into a national campaign. The threat to national culture and identity of the disappearance of wilderness was re-invoked by a new breed of ecologically-minded preservationists and outdoor recreation enthusiasts. In 1935 the Wilderness Society was formed 'for the purpose of fighting off the invasion of the wilderness and of stimulating . . . an appreciation of its multiform emotional, intellectual and scientific values' (in Nash 1967: 207). Strongly influenced by Aldo Leopold's ecological-cum-amenity philosophy of wilderness preservation and founded by Robert Marshall who had a profound belief in the psychological benefits of wilderness appreciation, the Wilderness Society quickly established itself as the leading voice of the wilderness preservation movement. It publicised threats to wilderness areas and promoted their use for environmentally sensitive forms of outdoor recreation. In this it was supported by the continued activities of the Sierra Club in the west as well as the work of nature and wildlife organisations like the Izaak Walton League and the National Audubon Society.

The ideas which led to the founding of the Wilderness Society anticipated by some thirty years the latest wave of wilderness preservation. Since the mid-sixties, the old arguments for protecting the last vestiges of untouched nature from the advances of civilisation have been re-worked by a new environmental movement which draws heavily on the ecological approach of Aldo Leopold. The Wilderness Society and the Sierra Club are, to a great extent, now part of a movement which also counts Friends of the Earth and Greenpeace amongst its members. Saving the wilderness is now synonymous with saving the planet. And so the wilderness movement has become more organised and more militant. New groups, sometimes under the umbrella of national organisations, sometimes independently, have sprung up around specific wilderness campaigns, using protest and demonstration as their principal weapons of publicity.

As I have suggested, the failure of an integrated countryside movement like Britain's to develop in North America can be traced to some fundamental differences in historical geography. Foremost amongst these, however, must surely be the preoccupation with wilderness which has dominated the North American approach to scenic and nature preservation for most of this century. It is a preoccupation which, despite the development of a countryside ideal which includes the settled rural landscape as well as the wilderness frontier, has diverted much of the continent's preservational attention away from the more immediate countryside. However, this does not mean that this countryside has been entirely ignored as a preservational issue. In fact, from a number of different perspectives, public interest in its protection has gradually increased. This has been fostered by three distinct movements: nature and wildlife conservation, historic preservation and farmland preservation. For the most part, these have functioned independently of each other.

Nature and wildlife

Much of the early public support for the protection of nature and wildlife was, of course, stimulated by the wilderness movement, which acted as a catalyst for a more general rise of concern over the disappearance of natural environments. From local naturalist societies, some of which date back, as in Britain, to the latter part of the last century, to powerful national organisations such as the National Audubon Society, an extensive nature and wildlife conservation network has developed which inevitably has focused much of its attention on the countryside. Founded in 1866, the National Audubon Society was the first national organisation to draw attention to the importance of habitat as well as species protection, and thus to stimulate interest in the preservation of specific elements of the rural landscape. In 1922 it was joined by the Izaak Walton League which added the somewhat ambiguous perspective of the hunting fraternity to the issue of habitat protection. Today, the National Audubon Society, with over 500 local chapters, leads the way in education and action programmes to protect wildlife and natural areas, while the Izaak Walton League promotes citizen involvement in local environmental protection. Both have large memberships and publish national magazines. Other nature groups have emerged during the last thirty years, most notably the Nature Conservancy, which, to some extent has assumed the responsibility for habitat protection with its programme of acquiring and protecting land that supports rare ecosystems (Stokes 1989).

Much as they do in Britain, North American nature and wildlife organisations exert their influence on more general countryside conservation through their involvement in local environmental conservation issues. Every anti-development campaign today, can be virtually certain of counting on the support of the nature groups, especially at the local level. Indeed it is the proliferation of local groups which has probably contributed more than anything to the growth in public support for the protection of rural landscapes. However, we must not ignore the increasingly important influence of the broader environmental lobby. While the wilderness remains the focus of its more controversial campaigns, the environmental lobby has done much to raise public concern over threats to the natural environment in the countrysides of most metropolitan regions.

Rural heritage

If wilderness preservation is one of the pillars of national identity, then historic preservation must surely be the other. The preservation of the built environments and artifacts of previous generations symbolises, as David Lowenthal (1976) has pointed out, the American need to create its own visible landscape history. At one level this has been directed at national monuments and architectural treasures, and in this sense has therefore been concerned with both urban and rural areas. But at another level, it has focussed on historic preservation as an aesthetic issue. One of the earliest attempts at heritage preservation on a national scale in the USA was the founding in 1901 of the American Scenic and Historic Preservation Society, which was modelled on the British National Trust with the aim of protecting 'notable features of the landscape in city or country made beautiful by nature or art . . . and generally to promote popular appreciation of the scenic beauties of America' (Runte 1979). Its descendant is the National Trust for Historic Preservation, which was established by congressional charter in 1949 as the leading voluntary organisation in the promotion of the protection of historic properties (Fedelchak and Wood 1988).

It was the work of the National Trust which was largely responsible for the passage of the National Historic Preservation Act 1966 which created the National Register of Historic Places under the control of the National Parks Service. In this respect, much of the Trust's activity and that of other organisations has tended to be concentrated on the preservation of single buildings and properties, predominantly in urban areas. In recent years, however, it has broadened its programmes to include small town and rural heritage preservation. This is consistent with a long-established strand of historic preservation which has focused on vernacular landscapes – on barns and farmhouses, fences and woodlots, hamlets and villages and other artifacts of traditional rural life. Small town and village preservation in particular has a long history at the local level especially in New England and other areas where the colonial legacy has left a significant mark on the landscape of rural settlements. Local historic preservation societies existed as early as the turn of the century and were influential in the evolution of restrictive zoning as the basis of the American system of local land use planning which emerged during the 1930s. Virtually every small town and village with any sense of its historical character now has its historical or heritage society which can tap into the resources of national organisations like the National Trust and the American Association for State and Local History, as well of a range of State and Federal grants and statutes (Stokes 1989). More significantly they have been incorporated into the political fabric of many rural communities, not only through their influence on local zoning laws and ordinances but also through their partnership with local historic district commissions. Local historic preservation also has strongly touristic overtones as communities recognise the commercial advantages both of revivalising the traditional architecture of main streets and creating whole settlements as living museums of rural life.

Farmland and open space

The growth of public support for the preservation of rural heritage in North America has been paralleled by the development of a vigorous farmland preservation movement. Concern about the impact of urbanisation on agricultural land, which had long been an issue in Britain and other parts of Europe, came late to North America. It was not until the late 1950s and early 1960s that the first voices began to be raised against uncontrolled urban sprawl. Neo-Malthusian inspired fears of global food shortages fostered a growing public awareness of the limits to earth, which was readily extended to concern over the future American and Canadian farmland resources. National, as well as global food security became linked to the issue of the conversion of high quality agricultural land to non-agricultural uses. That this quickly became a public issue is largely due to the extensive academic analysis of farmland conversion which, for the most part, portrayed a gloomy picture of its rate of loss and of its consequences for agricultural production. By the 1970s most American states and Canadian provinces had adopted policies to control farmland conversion (Furuseth and Pierce 1982). In the USA it became enough of a national issue for then President Carter to establish the National Agricultural Lands Commission in 1979. This produced data, which despite their dubious accuracy, further fixed the notion of the rapid 'disappearance' of farmland in the public mind.

The official response to the supposed problem of farmland conversion was accompanied, although not always prompted by the proliferation of farmland and foodland preservation groups. The only truly national organisation is the American Farmland Trust (AFT), which is committed to 'protecting America's farmland', through direct land acquisition, advice on policy development and acting as an information clearinghouse. Much of the AFT's work is channelled through the many state and local groups that have sprouted up all over the USA in the past decade or so. Similarly in Canada, where there is no national organisation, farmland preservation groups have proliferated at the provincial and local level. In Ontario, for example, there is now a fairly well-organised network of groups, including three province-wide organisations, which promote public awareness of farmland preservation issues and act as watchdogs over the implementation of the province's Foodland Guidelines.

The growth of this farmland preservation movement has been driven not only by concern over the loss of agricultural production capacity, but also by the sentiments about farming and farmscape which are at the very heart of the North American countryside ideal. Indeed many groups, especially those operating at the local level, are influenced by a mixture of nostalgic agrarianism, environmentalist concern for the health of the land and amenity valuation of open space and rural scenery. A good example is the Farmlands Conservation Project ('Save the Farmbelt') initiated in 1980 by the People for Open Space organisation in California's Bay Area. In answer to its own question, 'How do people in cities benefit by having farms and ranches near them?' the project proposed a broad amenity role for farmland as the basis of a permanent greenbelt.

> Only thus can the Bay Area's land continue to serve us: to produce a portion of our food; to supply high-quality water; to reduce the threat of floods and other hazards; to maintain the richness of plant and animal life; to give us room to wander and to breathe.
>
> (People for Open Space 1980: 7)

Viewed from this perspective it is not surprising that farmland preservation has become increasingly incorporated into the broader preservational objectives of exurban and rural communities. The past decade or so has seen a phenomenal rise in the level of grass-roots, community-based activism aimed at the general protection of rural environment and character. Wherever communities are faced with development pressures, preservational and conservational movements seem almost certain to form. In the USA one of the principal manifestations of this is the growth of land trusts. These are local, state or regional non-profit organisations which are 'directly involved in protecting land for its natural, recreational, scenic, historical or productive value' through direct acquisition, conservation easements and other forms of control over land development (Land Trust Alliance 1989). Initially established in New England in the late nineteenth century as the first wave of exurbanites sought to protect the Arcadian exclusivity of their communities and the natural landscapes of the surrounding countryside, land trusts have increased rapidly in the past twenty years. With nearly 900 trusts with over 700,000 members, it is, according to the trusts' umbrella organisation, the Land Trust Alliance, 'the fastest growing conservation movement today'. New England, with almost half of the nation's trusts, still dominates the movement, although the number of trusts in other regions continues to grow. One example is the Open Space programme in Boulder, Colorado which has preserved 17,000 acres since 1967 to protect mountain scenery and agricultural land as well as create hiking and biking trails (Elfring 1989).

While land trusts have tended to focus on the specific protection of natural areas, open space and farmland, their recent proliferation is increasingly associated with the integration of these concerns into a comprehensive programme of local community preservation and conservation. A particularly good example is that of the Berks County Conservancy in Pennsylvania. One of the more influential trusts, it works to protect the county's rural character by acquiring interests in farmland, historic sites and natural areas (Stokes 1989). It has also been involved in the selection of Oley, one of the county's small towns, as a demonstration community of the National Trust for Historic Preservation's recently announced historic countryside programme. The expansion of the National Trust beyond its more traditional mandate is a recognition of the convergence of land conservation and historic preservation issues in local rural communities. Since the establishment of its rural programme in 1979, the Trust has been involved in a number of projects to assist rural communities in preservation and restoration. But what is perhaps more important, it has mounted the first explicit proposals in North America for general countryside conservation (Plate 6.4). Inspired, no doubt, by the upsurge of grass-roots rural activism, the National Trust now appears to be casting itself as the guardian of the American countryside. In its recent volume, *Saving America's Countryside: A Guide to Rural Conservation*, the Trust outlines a broad set of strategies for local rural initiatives, within a framework which defines rural conservation as the integration of natural resource conservation, farmland retention, historic preservation and scenic protection (Stokes 1989).

The underlying emphasis of the National Trust's programme is on the 'historic countryside', and 'the protection and enhancement of America's rural heritage' (National Trust 1988: 7). This strikes a responsive chord in the increasingly exurbanite and environmentally-conscious society of much of small-town and rural America. Local rural preservationism appears to be a rapidly growing movement, especially where development pressures are strong. In Canada there are no national or even provincial organisations to match the National Trust, and little of the institutional momentum such as has been generated by the American land trust movement. Yet, the protection of rural character is no less of a rallying cry for communities threatened by development pressures than it is in the USA. Local historical societies and rural heritage groups can be found in most places which have acquired a residential amenity profile. Federal and provincial heritage programmes, together with local legislation provide an institutional framework within which these organisations can operate. A good example is the Local Conservation Advisory Committee in Ontario which can be established by local municipalities to create a framework through which heritage groups can influence local planning.

WHOSE MOVEMENT?

It has been estimated that in Britain there are about eighty national environmental organisations of which around fifty are directly concerned with rural and nature conservation (Lowe and Godyer 1983). To these can be added at least 1,200 local groups (Gregory 1976). The combined

Plate 6.4   The first campaign for preserving the American countryside: National Trust booklet, 1988

membership of national and local organisations (adjusting for those who belong to several organisations) is probably between two and three million, which is roughly 10 per cent of the adult population (Lowe and Godyer 1983). There is also a significantly larger percentage which expresses support through both donations and participation in special campaigns and programmes.

What these crude estimates do not reveal, of course, are the particular sources of this support. Recent research has tended to emphasise the inherent élitism of the movement on both sides of the Atlantic, pointing to the dominance of middle and upper-middle class membership of conservation organisations. (Eversley 1974; Buller and Lowe 1982). As we have seen, there has been an establishment tone to much of the countryside movement from the beginning and to a considerable extent it still relies on the intellectual and social élite for its leadership and draws its public support from the more affluent and educated end of the social spectrum. This is hardly surprising, given that it is this section of society which is most able to identify with and articulate the countryside ideal. To some extent this stems from altruistic motives; a conventional middle-class sense of responsibility to act for the common good and to preserve its countryside aesthetic in the national interest. Much of the middle-class concern for the countryside, however, stems from the transformation of the social composition of rural Britain over the past fifty or so years. As Howard Newby (1987) has so aptly put it, 'concern for the rural environment thus became a public issue because there was now residing in the countryside an affluent and articulate population . . . which was capable of mobilising itself politically' (227). These are the people, the established rural 'gentlefolk' and the newer middle-class exurbanites, who tend to dominate the local amenity groups through which the national organisations now channel much of their activity (Buller and Lowe 1982). Indeed it is at the local level that preservationist values seem to be most directly associated with the aesthetic values and vested interests of a social élite, and most at odds with the more utilitarian needs of the community.

Of course, the countryside movement in Britain has not been an entirely middle-class affair. In fact, it has long contained a populist strand. Much of the campaign for recreational access to the countryside, as we have seen, has been led by organisations with working-class roots. That the Ramblers' Association and the Youth Hostels Association have among the largest memberships today reflects the fact that they draw their support from across the social spectrum. In recent years even the more conservative organisations, like the National Trust and the CPRE have tried to broaden their appeal beyond their traditional base of support (Blunden and Curry 1985). To some extent this is associated with the unprecedented spread of middle-class values across British society over the past decade. Clearly, as this has occurred a growing percentage of the population will have identified with the countryside movement. Indeed it may no longer be particularly meaningful to analyse the movement in class terms. If indeed the countryside movement is now acquiring a more populist flavour, much of the credit must be given to the movement itself. In many respects, it has come to lead as much as follow British public opinion. It has done so through an increasingly organised and aggressive campaign aimed at promoting public appreciation of the countryside and public awareness of conservational and preservational issues. As a result the countryside movement, at least in Britain, has become an established feature of cultural life, with sophisticated media campaigns, an endless stream of publications, a growing selection of participatory conservational activities for everybody from school children to senior citizens and a veritable industry of fund raising consumerism. It receives considerable support from corporate-sponsored campaigns like the Shell Better Britain Campaign, and from the bevy of writers who keep up a steady lament for the future of the countryside (for example Shoard 1985). In short it plays a central role in maintaining the countryside ideal in the public mind.

The fact that the British countryside movement has developed into such a national institution is, however, not just a function of its cultural significance, but also of its considerable influence on the country's political agenda. Not only has it been effective in obtaining legislation on countryside access and conservation, it has also become an officially accepted participant in the countryside policy-making process, through formal links to government agencies like the Countryside Commission and the Nature Conservancy Council, and in its role as watchdog over policy implementation. In part, this is a function of the level of respectability that the major organisations of the movement have attained. After all, the very respectability of the countryside ideal itself has guaranteed that from the outset they would obtain the support of influential figures of the political establishment. But, from the mass trespasses of the thirties to today's sophisticated pressure tactics, it is also a product of vigorous lobbying and public campaigning. In fact, these tactics appear to be on the rise in recent years, as a new generation of countryside and environmental conservationists expose the inadequacies of existing legislation and decry the impact of Thatcherite land use planning policies.

In North America, as we have already seen, it is more difficult to discern a movement for the general protection of the countryside. However, nature, recreational and heritage organisations are increasingly active in the dissemination of conservational ideology to a wider public. This is particularly true of the environmental movement which is playing a central role in raising public consciousness about the importance of protecting non-urbanised environments. At the same time, the rapidly growing local rural preservation movement, while driven by community self-interest also promises to increase broader concern for rural heritage.

The most lasting impact of the countryside movements on both sides of the Atlantic will be on the countryside itself. Their principal aim, as I suggested at the beginning of this chapter, has been to politicise the countryside ideal; to make it part of the political agenda and therefore of the legislative framework. The result, in varying degrees, has been the growing incorporation of amenity ideology into the land use planning framework. The designation of recreational areas, the provision and control of access to open space, the protection of scenic landscapes and of rural heritage, the conservation of natural environments, and, in the case of Britain, the establishment of overall policies for countryside management, have all resulted from the campaigns of the various components of the countryside movements. The countryside ideal has thus become an institutionalised part of the landscape.

7

REFLECTIONS

The countryside ideal was forged in the rise of modern urban civilisation. This is the central theme of this book. However, as I said in the Introduction, it is not an ideal which can be dismissed as the trivial nostalgia of urbanites. On the contrary, its complexity and durability demand that it be recognised as a significant influence in the shaping of our cultural landscapes and our environmental values: hence the historical treatment of the subject. The affection for the countryside on the part of urban society, as we saw in the foregoing chapters, stretches back at least three centuries. That it has been nurtured by a variety of forces – social, intellectual, artistic, scientific and economic – means that it has been gradually woven into the very fabric of at least modern Anglo-American culture, if not of western culture as a whole.

It is through the complex interplay of these forces over a long period therefore that the countryside has acquired the symbolic status as the idyllic alternative to urban environments that it now enjoys. The first two chapters of this book explore the emergence of this attitude to the countryside from the profound changes which accompanied the urbanisation and industrialisation of England and the European occupation of the New World. It was not, however, simply a reaction to rapid change, and particularly rapid urbanisation, that elevated the countryside to preferential status. What must be stressed is that countryside idealism is a product of the very changes themselves. The growth of great commercial and industrial cities together with the capitalist penetration of the rural economy transformed urban–rural relationships and, in the process, sharpened the distinction between country and city. At the same time the rise of the bourgeoisie produced a social class which sought to establish its status both by separating itself from the urban working classes and by imitating the lifestyles of the country gentry. This growing middle class built country homes and Arcadian suburbs, took scenic tours, became amateur naturalists and landscape preservationists; in short turned to the countryside for pleasure and respectability. In this they were aided by their not inconsiderable wealth and by significant improvements in transportation technology which opened up the countryside for property investment and recreation.

Yet this new bourgeoisie also formed the core of an increasingly educated society. Their attitudes to nature and landscape, their reactions to urban society and environment, indeed their social and aesthetic values as a whole, were informed by the swirl of intellectual, scientific and artistic ideas which accompanied the rise of industrialism. The fundamental shifts in the understanding and appreciation of human relationships with nature, the intellectual reappraisal of industrialism, the philosophy of agrarianism, and the aesthetic and spiritual values of romanticism combined to place the countryside on an ideological pedestal. Art and literature served to popularise this ideology among the educated classes. Countryside idealism thus became a cultural movement. It was, as we have seen, a movement which emerged in the eighteenth century, but gathered its greatest momentum during the nineteenth century. Indeed it is fair to say that the countryside ideal is a product of the nineteenth century – of a thoroughly Victorian way of seeing the world – and that its twentieth-century expression is built largely on this cultural base.

That this can be seen as an Anglo-American cultural phenomenon is revealed in the transatlantic fusion of values that accompanied European settlement of the North American continent. While much of this had to do with the self-conscious transfer of English culture into the American landscape, it was by no means a one-way flow. The encounter with what European settlers, in their astonishing ignorance of the surviving impact of the indigenous population, perceived to be an untouched natural world, moulded a distinctively American romanticism which not only defined domestic attitudes to natural landscapes, but also found its way back across the Atlantic. At the same time, we can recognise the emergence of distinctively English and American versions of the countryside ideal, born of different social systems, agrarian ideals and landscape experiences. Yet what has also been emphasised in this book is the congruity of forces which have fashioned the idealisation of the countryside on both sides of the Atlantic: the middle-class reaction against urban environments, the flourishing of the armchair countryside, the development of planning and conservation ideas, and the translation of all of this into actual residential and recreational experiences.

Several writers, notably Coleman (1973) and Wiener (1981) have suggested that sentiment for the countryside has exhibited a cyclical pattern, waxing and waning with the varying condition of economy and society. Thus, according to this thesis, countryside idealism has grown during times of economic depression and insecurity when doubts about urban-industrial civilisation have deepened. Certainly we can look to the early and late nineteenth centuries, the First and Second World Wars and the years in between as periods when great uncertainty about and even rejection of the virtues of industrial progress were accompanied by outpourings of nostalgia for rural life and landscape. And we could argue that the growth of countryside idealism in our own times is associated with the social and economic failure of late twentieth-century urbanism.

However, we must be cautious of such a deterministic argument. The idealisation of the countryside has been sustained as much by the success of modern urbanism as it has by ambivalence towards it. Both the cultural dissemination and the physical expression of the ideal are products of an affluent and mobile urbanised society. Paradoxically the persistence of an ideal which seeks escape from the city has depended upon the very wealth and opportunity that an urban economy has created. At no time has this been more evident than today. While the renewed and expanding countryside idealism of the past two or three decades may well reflect the uncertainties surrounding the restructuring of industrialism, its real strength is drawn from the forces which this process has unleashed; forces which have released urban society from the imaginative and physical bonds of the city to an extent hitherto impossible.

And yet, lurking beneath the cultural fabrication of countryside idealism are the deeper anxieties about modern life. The suggestion of a congruity between turbulent times and the intensification of nostalgia for the countryside implies the persistence of a fundamental human need for connections with nature, land and community. Clearly the countryside ideal could not have been culturally nurtured in the absence of such abiding emotions. It implies a psychology which reflects, as Fraser Harrison (1982) has pointed out, the literal meaning of nostalgia; the sense of loss of home, of homesickness.

The countryside thus becomes a symbolic landscape because it conveys meanings which speak of the very associations which urbanism and modernism have broken, and which our nostalgia drives us to restore. Of course, as the whole thrust of this book has argued, the expression and reinforcement of these meanings, and hence the sustenance of the countryside ideal in general, has depended upon its cultural nurturing and social exploitation. It is these influences, rather than simply an elemental nostalgia which have fashioned the myths and images of the countryside and which have encouraged their translation into landscape experience. In the entrenchment of these myths and images into the way we perceive the countryside we have given it new meanings which have influenced not only our abstract values, but also how we use and treat its landscapes.

In the process, the countryside has been both appropriated and refashioned to match its symbolic associations. Large areas of rural land and wilderness have been appropriated for amenity use as country and vacation homes, in recreational facilities, in parks and trails. The countryside has also been appropriated for urban and suburban design and for the renewal of city landscapes. At the same time, it has been refashioned to conform to the specific stereotypes and agendas of the countryside ideal in the landscaping of country estates, the development of recreational areas and the preservation and restoration of rural heritage. Furthermore it has been appropriated by a broad conservation movement which sees the countryside as the last bastion of the natural order.

As we approach the end of the century, these processes appear to have gathered new momentum, driven by a post-industrial amalgam of urban decentralisation, flexible accumulation, information technology, amenity commodification and environmental consciousness. Thus it has significance for our living spaces in general, for it affects both country and city. However, given the theme of this book, it is the consequences for the countryside which most merit attention in these concluding remarks. What kind of countryside has emerged from the imposition of its idealisation upon its landscapes and its people? As far as the physical landscape is concerned the most obvious and extensive changes are associated with its use as a residential and recreational amenity. As we saw in Chapters 3 and 4 this has had a profound effect on the appearance of rural and natural landscapes. But because these landscape changes embody the essential meanings of the countryside ideal they also symbolise more fundamental changes to the social and economic fabric of rural areas.

The consequences of the spread of exurban and recreational population and land use into rural areas have been extensively studied. For the most part they are characterised in terms of conflict; land-use conflict between productive and amenity activities, economic conflict between development and conservation, social conflict between newcomers and rural folk. These conflicts are seen largely in class terms, with an emphasis on an adventitious middle-class urban population invading and transforming 'traditional' rural society. Over the longer span of the history of countryside idealism, as much of the discussion in this book suggests, this is a fairly accurate picture. The gentrification process, in particular, imposed a new social order on the countryside which, as Raymond Williams so persuasively argued, generally ignored the interests of rural folk (Williams 1973). Today it takes more subtle forms. Affluent newcomers inflate property values and oppose development thus reducing housing and employment opportunities for rural people. They favour the preservation of scenic quality over profitable farming and so impose restrictions on the productive use of land. They infiltrate the social and political establishment of rural communities and so change the local agenda to suit their gentrified interests. Meanwhile, the more general recreational invasion of rural areas, in disrupting traditional uses of natural resources, has undermined rural lifestyles, especially amongst the indigenous peoples of North America. And it too has brought new values and demands to rural communities.

Yet to characterise this solely in terms of a conflict between urban, middle-class amenity seekers and a powerless rural population is an over-simplification of the impact of countryside idealism on rural areas. In the first place, conflict beween newcomers and established rural inhabitants does not always occur along class lines. Some of the most serious conflicts involve the two principal landowners in many rural areas, exurbanites and farmers, in which the incompatibility of amenity conservation and modern agricultural practices are the central issue. Furthermore it is probably no longer realistic to divide rural society into an urban middle class and a rural class with divergent attitudes towards the countryside. The vast majority of rural inhabitants today are non-farm folk with a diversity of backgrounds, both rural and urban, but with an increasingly convergent set of middle-class values, foremost among which we will find an amenity perspective. A second reason for questioning the urban–rural conflict model, is the persistence of conflict between amenity-users of the countryside. This is manifested in the long-standing dispute over the private ownership and public use of the countryside, over the whole question, that is, of its general accessibility. It is apparent also in the growing divergence between conservational and consumptive approaches to countryside amenities. Finally, the development of the countryside as both a public and a private amenity can bring some benefits to rural communities. It can raise the level of public services and stimulate the local economy through the diversification of commercial activities. It can lead to the revival of rural traditions and the garnering of interest in the preservation of local heritage. In short it can be the cultural and economic salvation of communities which would otherwise be submerged or marginalised by the forces of urbanisation.

Whatever the complexities of the impact of the countryside ideal on rural areas, the clear fact remains that it has become firmly established as a heritage and amenity landscape both in the public mind and in the lifestyles of a significant proportion of the population. The character of the countryside will therefore continue to be strongly influenced by an agenda set by a dominant rural non-farm population and supported by a broad environmental and heritage conservation movement. Given the diverse interpretations of the countryside ideal, this is not a unified agenda. But it is a far-reaching one, bolstered as it is by public policies and community initiatives. The British, or at least the English countryside, is a symbol of national identity and its preservation is somewhat of a national obsession. Although it seems unlikely to achieve quite this status in North America, on both sides of the Atlantic the enjoyment and protection of the countryside in a spirit consistent with its long-standing idealisation has become an important political issue. This may bode well for natural environment and cultural heritage. The countryside movement, in which exurbanites and recreational groups play a leading role, has become a force to be reckoned with in the battle to protect nature, wildlife, and scenic and cultural landscapes from the ravages of industrial agriculture and urban sprawl. These values have also extended to the improvement of urban environments.

However, while the modern countryside ideal has contributed to the general increase in environmental awareness and has resulted in the preservation of significant natural areas and cultural features, the sustainability of the settlement patterns which accompany it is more questionable. The spread of residential and recreational population across the countrysides of metropolitan regions and beyond brings with it all the environmental, economic and social costs of a decentralised settlement form. Despite the recurrent introduction of planning controls to deal with this in Britain and because of the paucity of such controls in North America, in many regions this is what constitutes the countryside. And it is this sprawling, dynamic and often disorganised 'rural' landscape, into which those with the resources to do so are moving in increasing numbers. The countryside, from its diluted suburban edge to its distant corners, has become for many the preferred place to live and play. This is where much of the complacent majority, as Galbraith (1992) has recently termed it, lives out its comfortable existence, increasingly served by decentralised jobs and facilities which permit it to avoid the city altogether.

The inner city and the old suburbs meanwhile, apart from a few exclusive enclaves, continue to degenerate into the squalor of poverty and deprivation, worlds apart from the countryside around. These are not just spatial distinctions but ones which go to the very heart of the social, economic and ethnic divisions of society. For the poor black and hispanic populations of American and the ethnically diverse immigrant populations of British cities, for the single mothers, the unemployed youth and the ageing poor that cluster in urban ghettos, the countryside might just as well be another planet.

And so the contemporary expression of the countryside ideal promises to perpetuate and solidify the social and economic disparities between country and city, or more precisely, between the old city cores and the new decentralised suburban, exurban and recreational landscapes beyond. It is in this new countryside that political and economic power may well concentrate, revitalising rural communities in the process, but also threatening the very amenities that countryside idealism seeks to enjoy.

SUGGESTED FURTHER
READING

Although most of the items listed below have already been referred to in the text, they are summarised here in order to highlight for readers those sources which can best form the basis for further exploration of the countryside ideal.

GENERAL BACKGROUND

Clearly the single most important other book to deal with the countryside ideal is Raymond William's *Country and the City* (1973). Although it explores the subject from the perspective of English literature and of the English class system, its discussion of the cultural origins of contrasting attitudes to country and city make it essential reading. Again with an English perspective, Fraser Harrison's *Strange Land, The Countryside: Myth and Reality* (1982) offers a critical appraisal of rural nostalgia. Other general discussions of the idealisation of countryside are limited to single chapters in books. Chapter 8 in Tuan's *Topophilia* (1974) and Chapters 2 and 4 in *Imagined Country* by John Short (1991) offer interesting interpretations. For an American overview, Schmitt's *Back to Nature: The Arcadian Myth in Urban America* (1969) is also a valuable source. Tuan's and Short's books are part of a broader body of literature which examines the cultural origins of landscape ideals. Among the most significant are Relph's *Rational Landscapes and Humanistic Geography* (1981), Cosgrove's *Social Formation and Symbolic Landscape* (1984) and Donald Meinig's edited volume, *The Interpretation of Ordinary Landscapes* (1979). For a thoroughly stimulating exploration of the meanings of the North American landscape, Alexander Wilson's *The Culture of Nature* (1990), is highly recommended.

THE MAKING OF AN IDEAL

Several key sources were used in the discussion of the historical and ideological formation of the ideal, and these all merit a full reading. Especially recommended are Coleman, *The Idea of the City in Nineteenth*

Century Britain (1973); Glacken, *Traces on the Rhodian Shore* (1967); Huth, *Nature and the American: Three Centuries of Changing Attitudes* (1957); Meinig, *The Shaping of America* (1986); Nash, *Wilderness and the American Mind* (1967); Newby, *Country Life: A Social History of Rural England* (1987); Shi, *The Simple Life: Plain Living and High Thinking in American Culture* (1985); Weiner, *English Culture and the Decline of the Industrial Spirit* (1981); and Morton and Lucia White's *The Intellectual Versus the City* (1962). Also well worth reading in the original are William Morris's *News From Nowhere* (1891) and David Bellamy's *Looking Backward* (1888).

THE ARMCHAIR COUNTRYSIDE

There is a wealth of material on the literary and artistic treatment of nature, rural life and landscape, but most deal with specific genres and are very 'literary' in approach. In addition to Raymond Williams who is mentioned above, among the more general discussions which deserve further attention are Finch and Elder, *The Norton Book of Nature Writing* (1990); Keith, *The Rural Tradition* (1974); Mingay, *The Rural Idyll* (1989); Carpenter, *Secret Gardens: A Study of the Golden Age of Children's Literature* (1985); and Clough, *The Necessary Earth: Nature and Solitude in American Literature* (1964).

A PLACE IN THE COUNTRY

On the English country house phenomenon I especially recommend Clemenson, *English Country Houses and Landed Estates* (1982); and Girouard, *Life in the English Country House* (1978). Newton's compendious survey *Design on the Land* (1971) examines both the English and the American country estate tradition, but for by far the best study of the American country house, read Clive Aslet's magnificently illustrated book, *The American Country House* (1990).

Exurbia has been extensively studied, but of particular note are Spectorsky's original thesis, *The Exurbanites* (1955); Ronald Blyth's seminal survey of an exurbanising Suffolk village, *Akenfield* (1969); and Kenneth Jackson's *Crabgrass Frontier* (1985). Dorst's post-modernist interpretation of an exclusive Pennsylvanian exurban community, *The Written Suburb* (1990) also merits special attention. For less original but otherwise informative treatments read Bryant *et. al. The City's Countryside* (1982) and my own *Rural Settlement in an Urban World* (1982). The most interesting analysis of exurbia, however, is John Punter's 'Urbanites in the Countryside' (1974) which unfortunately is an unpublished Ph.D. thesis and therefore not easy to obtain.

On the back-to-the-land movement I recommend two retrospective

213

studies, Berger's *The Survival of a Counterculture* (1981) and Marsh's *Back to the Land* (1984). But for the original inspiration, everyone, of course, should read Thoreau's *Walden* (1854).

THE PEOPLE'S PLAYGROUND

Although there are plenty of small case studies of countryside recreation, good conceptual analyses of the subject are few and far between. Ian Ousby's *The Englishman's England: Taste, Travel and the Rise of Tourism* (1990) provides a fascinating examination of the early years of English countryside touring, while Hans Huth's, *Nature and the American* (1959) and Schmitt's *Back to Nature* (1969) still offer the best account of nineteenth-century American rural and wilderness recreation. The most interesting studies of modern countryside recreation include Coppock and Duffield, *Recreation in the Countryside* (1975); Simmons, *Rural Recreation in the Industrial World* (1975). Beyond these there is little to recommend apart from works on the management of recreational resources such as MacEwan's *National Parks: Conservation or Cosmetics?* (1982).

THE COUNTRY IN THE CITY

On the subject of parks, Chadwick's *The Park and the Town, Public Landscape in the 19th and 20th Century* (1966) is the most comprehensive source, while Schuyler's *The New Urban Landscape* (1986) includes several sections on the urban parks movement in the USA. On the garden city and suburb, the latter is again one of the seminal works, as well as Creese, *The Search for Environment: The Garden City, Before and After* (1966); Fishman, *Urban Utopias in the Twentieth Century* (1977) and Hall, *Cities of Tomorrow: An Intellectual History* (1988). On Arcadian suburbia in general, Edwards, *The Design of Suburbia* (1981) offers the best account of the British scene, while Jackson's *Crabgrass Frontier* (1985) presents an excellent analysis of the suburbanisation of America. On the subject of urban greening, the classics are Garrett Eckbo's *The Landscape We See* (1969) and Ian McHarg's *Design With Nature* (1969). However, Michael Hough's *City Form and Natural Process* (1984) and Nicholson-Lord's *The Greening of Cities* (1987) represent the more current thinking on the incorporation of nature into urban design.

THE COUNTRYSIDE MOVEMENT

The literature on this subject is dominated by the British perspective. Among the few American sources, Nash's *Wilderness and the American Mind* (1967) and Runte's *National Parks: The American Experience* (1979) provide comprehensive accounts of the wilderness preservation movement.

214

Of the many studies of the British movement, I recommend Chapter 4 in Blunden and Curry, *The Changing Countryside* (1985); Hill's *Freedom to Roam* (1980) and Stephenson's *Forbidden Land* (1989), which together furnish us with the complete story of the countryside access movement; Sheail's *Rural Conservation in Inter-War Britain* (1981); Lowe and Godyer's excellent *Environmental Groups in Politics* (1983) and a very recent book by David Evans, *A History of Nature Conservation in Britain* (1992). Of course no list of books on British countryside conservation would be complete without the inclusion of Marion Shoard's attack on the forces of rural landscape degradation, *The Theft of the Countryside* (1985).

BIBLIOGRAPHY

Abercrombie, P. (1926) *The Preservation of Rural England*, Liverpool: Liverpool University Press.
—— (1945) *The Greater London Plan, 1944*, London: HMSO
Armytage, W. (1961) *Heavens Below; Utopian Experiments in England, 1560–1960*, Toronto: University of Toronto Press.
Aslet, C. (1990) *The American Country House*, New Haven: Yale University Press.
Baldwin, S. (1926) *On England, and Other Addresses*, London: Philip Allan.
Balsdon, J. (1969) *Life and Leisure in Ancient Rome*, London: The Bodley Head.
Bellamy, D. (1888) *Looking Backward*, New York: New American Library, 1960 edition.
- Benningfield, G. (1983) *Hardy Country*: London, Penguin.
Beresford, J.T. (1966) *Land and People*, London: Leonard Hill.
Berger, B. (1981) *The Survival of a Counterculture*, Berkeley: University of California Press.
Betjeman, J. (1982) 'Introduction' in T. Evans and C. Green (eds) *English Cottages*, London: Weidenfeld and Nicholson.
Binford, H.C. (1985) *The First Suburb: Residential Communities on the Boston Periphery, 1815–1860*, Chicago: University of Chicago Press.
Blount, M. (1974) *Animal Land*, New York: Avon Books.
- Blunden, J. and Curry, N. (1985) *The Changing Countryside*, London: The Open University/Croom Helm.
—— (eds) (1990) *A People's Charter?*, London: HMSO for the Countryside Commission.
Blyth, R. (1969) *Akenfield*, London: Penguin.
Borsodi, R. (1933) *Flight From the City, An Experiment in Creative Living on the Land*, New York: Harper.
- Bowers, W. (1974) *The Country Life Movement in America, 1900–1920*, Port Washington, NY: Kennikat Press.
British Travel Authority (1967) *Pilot National Recreation Survey*, Keele: Keele University.
- Brown, D. and Wardwell, J. (1984) *New Directions in Urban-Rural Migration*, New York: Academic Press.
Bryant, C., Russwurm, L. and McClellan, A. (1982) *The City's Countryside*, London: Longman.
Buller, H. and Lowe, P. (1982) 'Politics and Class in Rural Preservation: A Study of the Suffolk Preservation Society', in M. Moseley (ed.) *Power, Planning and People in Rural East Anglia*, Norwich: University of East Anglia.

Bunce, M. (1982) *Rural Settlement in an Urban World*, London: Croom Helm.

Burnett, J. (1978) *A Social History of Housing, 1815–1870*, Newton Abbott: David and Charles.

Burton, R. J. C. (1973) *The Recreational Carrying Capacity of the Countryside*, Keele: University of Keele.

Butlin, R. (1982) *The Transformation of Rural England c. 1580–1880*, London: Oxford University Press.

Carpenter, H. (1985) *Secret Gardens: A Study of the Golden Age of Children's Literature*, London: Allen and Unwin.

Cavaliero, G. (1977) *The Rural Tradition in the English Novel, 1900–1939*, London: Macmillan.

Chadwick, G. F. (1966) *The Park and the Town, Public Landscape in the 19th and 20th Century*, London: Architectural Press.

Chambers, J. and Mingay, G. (1966) *The Agricultural Revolution, 1750–1880*, London: Batsford.

Champion, A., Fielding, A. and Keeble, D. (1989) 'Counter-Urbanization in Europe', *Geographical Journal*, 155: 52–80.

Christensen, C. (1978) *The American Garden City and the New Towns Movement*, Ann Arbor: UMI Research Press.

Civic Trust (1990) *Amenity in Action: The Civic Trust Handbook of Amenity Initiatives*, London: Shell Better Britain Campaign and Civic Trust.

Clay, G. (1973) *Close Up; How to Read the American City*, New York: Praeger.

Clemenson, H. (1982) *English Country Houses and Landed Estates*, London: Croom Helm.

Cloke, P.J. and Park, C. (1984) *Rural Resource Management*, London: Croom Helm.

Clough, Wilson O. (1964) *The Necessary Earth: Nature and Solitude in American Literature*, Austin: University of Texas Press.

Cobbett, W. (1912) *Rural Rides*, London: J. M. Dent.

Coleman, R. (ed.) (1973) *The Idea of the City in Nineteenth Century Britain*, London: Routledge and Kegan Paul.

Connell, J. (1978) *The End of Tradition: Country Life in Central Surrey*, London: Routledge and Kegan Paul.

Cooper, J. (1978) 'In Search of Agrarian Capitalism', *Past and Present*, 80: 37–47.

Coppock, J. (1977) *Second Homes; Curse or Blessing?*, Oxford: Pergamon.

Coppock, J. and Duffield, B. (1975) *Recreation in the Countryside*, London: Macmillan.

Coppock, J. and Prince, H. (eds) (1964) *Greater London*, London: Faber and Faber.

Cordell, H., and Hendee, J. (1982) *Renewable Resources Recreation in the United States*, Washington, D.C.: American Forestry Association.

Cosgrove, D. (1984) *Social Formation and Symbolic Landscape*, London: Croom Helm.

Council For National Parks (1986) *50 Years of National Parks*, London.

Council for the Preservation of Rural England (1927) *Annual Report*, London.

Countryside Commission (1985a) *National Countryside Recreation Survey: 1984*, Cheltenham.

—— (1985b) 'Kids in the Countryside', *Countryside Commission News*, 16: 4.

—— (1985c) 'City sanctuaries', *Countryside Commission News*, 17: 5.

Cox. G. (1976) 'The Hardy Industry', in M. Drabble (ed.) *The Genius of Thomas Hardy*, New York: Knopf.

Creese, W. L. (1966) *The Search for Environment: The Garden City, Before and After*, Boston: Yale University Press.

Cunningham, H. (1980) *Leisure in the Industrial Revolution c. 1780-c.1880*, New York: St. Martin's Press.

Dahms, F. (1988) *The Heart of the Country*, Toronto: Deneau.

Darley, G. (1975) *Villages of Vision*, London: Architectural Press.

Dennis, R. (1984) *English Industrial Cities of the Nineteenth Century: A Social Geography*, Cambridge: Cambridge University Press.

Dickens, C. (1854) *Hard Times*, New York: New American Library, 1961 edition.

Ditchfield, P. H. (1898) *The Charm of the English Village*, London: Bracken Books, 1985 edition.

Dorst, J. (1990) *The Written Suburb*, Philadelphia: University of Pennsylvania Press.

Dower, M. (1984) 'Green Towns and Cities' *The Planner*, July, 1–2.

Downing, A. (1851) *The Architecture of Country Houses*, New York: D. Appleton.

Duany, A. (1991) *Towns and Town-Making Principles*, New York: Harvard Graduate School of Design.

Duffield, B. and Owen, M. (1970) *A Geographical Appraisal of Countryside Recreation in Lanark*, Edinburgh: University of Edindurgh.

Duncan, J. (1973) 'Landscape Taste as a Symbol of Group Identity, a Westchester County Village', *Geographical Review*, 63: 336.

Dunford, M. and Perrins, D. (1983) *The Arena of Capital*, London: Macmillan.

Eckbo, G. (1969) *The Landscape We See*, New York: McGraw-Hill.

Ecksteins, M. (1990) *Rites of Spring*, Toronto: Lester and Orpen Dennys.

Edwards, A. M. (1981 *The Design of Suburbia*, London: Pembridge Press.

Elder-Duncan, J. (1912) *Country Cottages and Week-End Homes*, London: Cassell.

Elfring, C. (1989) 'Preserving Land Through Local Land Trusts,'*BioScience*, 39: 71–3.

Elliott, E. (ed.) (1991) *The Columbia History of the American Novel*, New York: Columbia University Press.

Ellis, A. (1968) *A History of Children's Reading and Literature*, Oxford: Pergamon.

Emery, M. (1986) *Promoting Nature in Cities and Towns*, London: Croom Helm.

Eustice, A. (1979) *Thomas Hardy: Landscape of the Mind*, New York: St. Martin's Press.

Evans, D. (1992) *A History of Nature Conservation in Britain*, London: Routledge.

Eversley, D. (1974) 'Conservation for the Minority,' *Built Environments*, 3: 14–15.

Fedelchak, M. and Wood, B. (1988) *Protecting America's Countryside*, Washington, D.C.: National Trust for Historic Preservation.

Finch, R. and Elder, J. (eds) (1990) *The Norton Book of Nature Writing*, New York: W. W. Norton.

Finsterbusch, K. (1980) *Understanding Social Impacts*, London: Sage Books.

Fishman, R. (1977) *Urban Utopias in the Twentieth Century*, New York: Basic Books.

—— (1987) *Bourgeois Utopias*, New York: Basic Books.

Fitzgerald, F. S. (1925) *The Great Gatsby*, New York: Bantam Books.

Flanagan, J. (1986) 'Green Vision', *Architectural Journal*, 5 Feb.: 48–50.

Foerster. N. (1950) *Nature in American Literature*, New York: Russell and Russell.

Fogarty, R. (1972) *Dictionary of American Communal and Utopian History*, Westport: Greenwood Press.

Fox, S. (1981) *John Muir and His Legacy: The American Conservation Movement*, Boston: Little, Brown and Co.

Frye, N. (1967) *The Romantic Myth*, New York: Random House.

Furst, L. (1971) *Romanticism*, London: Methuen.

218

Furuseth, O. and Pierce, J. (1982) *Agricultural Land in an Urban Society*, Washington, D.C.: Association of American Geographers.

Galbraith, J. K. (1992) *The Culture of Contentment*, Boston: Houghton Mifflin.

Gans, H. (1967) *The Levittowners: Ways of Life and Politics in a New Suburban Community*, New York: Pantheon.

—— (1982) *The Urban Villagers: Group and Class in the Life of Italian-Americans*, New York: Free Press.

Geddes, P. (1915) *Cities in Evolution*, London: Williams.

Gilg, A. (1986) *An Introduction to Rural Geography*, London: Arnold.

Girouard, M. (1978) *Life in the English Country House*, New Haven: Yale University Press.

Gitlin, T. (1983) *Inside Prime Time*, New York: Pantheon Books.

Glaab, C. and Brown, A. T. (1976) *A History of Urban America*, New York: Macmillan.

Glacken, C. (1967) *Traces on the Rhodian Shore*, Berkeley: University of California Press.

Good, H. E. (1967) *Black Swamp Farm*, Columbus: Ohio State University Press.

Gottmann, J. (1963) *Megalopolis*, New York: Twentieth Century Fund.

Graves, J. (1980) *From a Limestone Edge*, New York: Alfred A. Knopf.

Green, F. (1912) *The Awakening of England*, London: Nelson.

Gregory, R. (1976) 'The Voluntary Amenity Movement,' in A. MacEwan (ed.) *Future Landscapes*, London: Chatto and Windus.

Haggard, H. R. (1905) *The Poor and the Land*, London: Longmans, Green and Co.

Hahn, S. and Prude, J. (eds) (1985) *The Countryside in the Age of Capitalist Transformation*, Chapel Hill: University of North Carolina Press.

Hall, C. (1976) 'The Amenity Movement,' in Gill, C. (ed.) *The Countryman's Britain*, Newton Abbott: David and Charles.

Hall, M. (1977) 'The Park at the End of the Trolley', *Landscape*, 22: 11–18.

Hall, P. (1988) *Cities of Tomorrow: An Intellectual History*, London: Blackwell.

Hammitt, W. E. and Cole, D. N. (1987) *Wildland Recreation: Ecology and Management*, New York: Wiley.

Hardy, D. (1979) *Alternative Communities in Nineteenth Century England*, London: Longman.

Harris, R. C. (1966) *The Seigneurial System in Early Canada*, Toronto: University of Toronto Press.

—— (1977) 'The Simplification of Europe Overseas', *Annals, Association of American Geographers*, 67: 469–83.

Harrison, C., Limb, M. and Burgess, J. (1987) 'Nature in the City – Popular Values for a Living World,' *Journal of Environmental Management*, 25: 347–62.

Harrison, F. (1982) *Strange Land*, London: Sidgwick and Jackson.

Hart, J. (1950) *The Popular Book: A History of America's Literary Taste*, New York: Oxford University Press.

Hawkins, D. (1983) *Hardy's Wessex*, London: Macmillan.

Hayden, D. (1984) *Redesigning the American Dream*, New York: W. W. Norton.

Healy, R. and Short, J. (1983) *The Market for Rural Land*, Washington D.C.: Conservation Foundation.

Hill, H. (1980) *Freedom to Roam: The Struggle for Access to Britain's Moors and Mountains*, Ashbourne, Derbyshire: Moorland Publishing.

Hobsbawm, E. (1968) *Industry and Empire*, London: Penguin.

Hofstadter, R. (1966) *The Age of Reform*, New York: Alfred A. Knopf.

219

BIBLIOGRAPHY

Horne, D. (1986) *The Public Culture; The Triumph of Industrialism*, London: Pluto.

Hough, M. (1984) *City Form and Natural Process: Towards a New Urban Vernacular*, New York: Van Nostrand Reinhold.

Howard, E. (1898) *Garden Cities of Tomorrow*, London: Faber and Faber, 1946 edition.

Hugo-Brunt, M. (1967) 'Downing and the English Landscape Tradition', 'Introduction' in A. Downing, *Cottage Residences, Rural Architecture and Landscape Gardening*, New York: Library of Victorian Culture.

Humphreys, E. (1964) *The Augustan World*, London: Methuen.

Hunter, S. (1984) *Victorian Idyllic Fiction*, London: Macmillan.

Hussey, C. (1967) *The Picturesque: Studies of a Point of View*, London: F. Cass.

Huth, H. (1957) *Nature and the American: Three Centuries of Changing Attitudes*, Berkeley: University of California Press.

Hyams, E. (1971) *Capability Brown and Humphrey Repton*, London: J. M. Dent.

Jackson, E. L. (1986) 'Outdoor Recreation and Environmental Attitudes', *Leisure Studies*, 5: 1–24.

Jackson, K. (1985) *Crabgrass Frontier*, New York: Oxford University Press.

Jacobs, J. (1961) *The Death and Life of Great American Cities*, New York: Random House.

James, L. (1989) 'Landscape in Nineteenth Century Literature', in G. E. Mingay, (ed.) *The Rural Idyll*, London: Routledge.

Jellicoe, G. and Jellicoe, S. (1975) *The Landscape of Man, Shaping the Environment from Prehistory to the Present Day*, London: Thames and Hudson.

Johansen, H. and Fuguitt, G. (1984) *The Changing Rural Village in America*, Cambridge, Mass.: Ballinger.

Johnson, E. D. H. (ed.) (1966), *The Poetry of Earth: A Collection of English Nature Writings*, New York: Atheneum.

Jones, E. L. (1968) 'The Agricultural Origins of Industry,' *Past and Present*, 40: 58–71.

Jones, H. R. (1965) *John Muir and the Sierra Club: The Battle for Yosemite*, San Francisco: Sierra Club.

Kaplan, R. (1984) 'Wilderness Perception and Psychological Benefits: An Analysis of a Continuing Program,' *Leisure Studies*, 6: 271–90.

Karolides, N. J. (1967) *The Pioneer in the American Novel, 1900–1950*, Norman: University of Oklahoma Press.

Kazin, A. (1988) *A Writer's America*, New York: Alfred A. Knopf.

Keith, W. (1974) *The Rural Tradition*, Toronto: University of Toronto Press.

—— (1980) *The Poetry of Nature*, Toronto: University of Toronto Press.

—— (1988) *Regions of the Imagination: The Development of British Rural Fiction*, Toronto: University of Toronto Press.

Kellett, J. (1969) *The Impact of Railways on Victorian Cities*, London: Routledge and Kegan Paul.

Koster, D. (1975) *Transcendentalism in America*, Boston: Twayne.

Kraus, R. (1984) *Recreation and Leisure in Modern Society*, Glenview, Ill.: Scott Foresman.

Kristol, I. (1970) 'Urban Civilisation and its Discontents', *Commentary*, 50: 29–36.

Land Trust Alliance (1989) *National Directory of Conservation Land Trusts*, Washington, DC.

Laurie, I. C. (1979) *Nature in Cities*, London: Wiley.

Lavery, P. (1971) *Recreational Geography*, London: David and Charles.

Leavis, Q. (1932) *Fiction and the Reading Public*, London: Chatto and Windus.

Lemon, J. (1972) *The Best Poor Man's Country*, Baltimore: Johns Hopkins University Press.

—— (1984) 'Spatial Order: Households in Local Communities and Regions', in J. Greene and J. Pole (eds) *Colonial British America*, Baltimore: Johns Hopkins University Press.

Leopold, Aldo (1949) *A Sand County Almanac*, New York: Oxford University Press.

Lerner, L. and Holmstrom, J. (1968) *Thomas Hardy and His Readers*, London: The Bodley Head.

Lewis, P (1982) 'The Galactic Metropolis', in G. Rutherford Platt and G. Demko (eds) *Beyond the Urban Fringe*, Washington, DC: Association of American Geographers.

Lewis, R. and Maude, A. (1949) *The English Middle Classes*, London: Phoenix House.

Lora, R. (1971) *Conservative Minds in America*, New York: Rand McNally.

Lowe, P. and Godyer, J. (1983) *Environmental Groups in Politics*, London: Allen and Unwin.

Lowenthal, D. (1976) 'The Place of the Past in the American Landscape', in D. Lowenthal and M. Bowden (eds) *Geographies of the Mind*, New York: Oxford University Press.

Lowenthal, D. and Prince, H. (1965) 'English Landscape Tastes', *Geographical Review*, 55: 187–222.

Mabey, R. (1983) 'Introduction' in R. Jefferies *Landscape With Figures: An Anthology of Richard Jefferies's Prose*, London: Penguin Books.

MacEwan, A. (1982) *National Parks: Conservation or Cosmetics?* London: Allen and Unwin.

McHarg, I. (1969) *Design With Nature*, Garden City, NJ: Natural History Press.

McMullen, J. (1984) *My Small Country Living*, London: Allen and Unwin.

Maguire, J. H. (1991) 'Fiction of the West', in E. Elliott (ed.) *The Columbia History of the American Novel*, New York: Columbia University Press.

Marchand, R. (1985) *Advertising the American Dream: Making Way for Modernity, 1920–40*, Berkeley: University of California Press.

Marinelli, P. V. (1971) *Pastoral*, London: Methuen.

Marsh, J. (1984) *Back to the Land*, London: Quartet Books.

Marshall, D. (1968) *Dr. Johnson's London*, London: Wiley.

Marx, L. (1964) *The Machine in the Garden*, New York: Oxford University Press.

Massingham, H. (1945) 'Introduction' in F. Thompson *Lark Rise to Candleford*, London: Oxford University Press.

Matthias, P. (1969) *The First Industrial Nation: An Economic History of Britain, 1700–1914*, London: Methuen.

Meinig, D. W. (1979) 'Symbolic Landscapes', in D. W. Meinig (ed.) *The Interpretation of Ordinary Landscapes*, New York: Oxford University Press.

Meinig, D. (1986) *The Shaping of America*, New York: Yale University Press.

Melville, K. (1972) *Communes in the Counter Culture*, New York: William Morrow.

Merrington, J. (1976) 'Town and Country in the Transition to Capitalism', in R. Sweezy *et al.* *The Transition from Feudalism to Capitalism*, London: N.L.B.

Meyer, R. W. (1965) *The Middle Western Farm Novel in the Twentieth Century*, Lincoln: University of Nebraska Press.

Mingay, G. E. (ed.) (1989) *The Rural Idyll*, London: Routledge.

Monkonnen, E. (1988) *America Becomes Urban; the Development of U.S. Cities and Towns, 1780–1980*, Berkeley: University of California Press.

BIBLIOGRAPHY

Morris, W. (1891) *News From Nowhere*, London: Routledge and Kegan Paul, 1970 edition.

Muir, J. (1890), 'The Treasures of the Yosemite', *Century Magazine* 40: 483; and 'Features of the Proposed Yosemite National Park', *Century Magazine* 41: 666–7.

Muir, R. (1981) *The English Village*, London: Thames and Hudson.

Muller, P. (1976) *The Outer City: Geographical Consequences of the Urbanization of the Suburbs*, Washington, DC: Association of American Geographers.

Mumford, L. (1926) *The Golden Day*, New York: Boris and Liveright.

—— (1951) *The Conduct of Life*, New York: Harcourt, Brace.

—— (1961) *The City in History*, New York: Harcourt, Brace and World.

Nash, R. (1967) *Wilderness and the American Mind*, New Haven: Yale University Press.

—— (1968) *The American Environment: Readings in the History of Conservation*, Reading, Mass.: Addison-Wesley.

National Trust for Historic Preservation (1988) *Protecting America's Countryside*, Washington, DC.

Newby, H. (1987) *Country Life: A Social History of Rural England*, Totowa, N.: Barnes and Noble.

Newton, N. (1971) *Design on the Land, The Development of Landscape Architecture*, Cambridge, Mass.: Belknap Press.

Nicholson, M. (1987) *The New Environmental Age*, Cambridge: Cambridge University Press.

Nicholson-Lord, D. (1987) *The Greening of Cities*, London: Routledge and Kegan Paul.

Novak, Barbara (1980) *Nature and Culture: American Landscape Painting, 1825–1875*, New York: Oxford University Press.

Oliver, P., Davis, I. and Bentley, I. (1981) *Dunroamin: The Suburban Semi and its Enemies*, London: Barrie and Jenkins.

Olmsted, F. L. (1870) 'Public Parks and the Enlargement of Towns', in *Civilizing American Cities: A Selection of Frederick Law Olmsted's Writings on City Landscapes*, Cambridge, Mass.: M.I.T. Press, 1971 edition.

Osborn, F. and Whittick, A. (1969) *The New Towns: The Answer to Megalopolis*, London: L. Hill.

Ousby, I. (1990) *The Englishman's England: Taste, Travel and the Rise of Tourism*, Cambridge: Cambridge University Press.

Pacione, M. (1985) *Rural Geography*, London: Harper and Row.

Pahl, R. (1965) *Urbs in Rure*, London: Geography Paper 2, London School of Economics.

Patel, D. I. (1980) *Exurbs: Residential Development in the Countryside*, Washington DC: University Press of America.

Paterson, R. (1989) 'Creating the Packaged Suburb: The Evolution of Planning and Business Practices in the Early Canadian Land Development Industry, 1900–1914', in B. Kelly (ed.) *Suburbia Re-examined*, Westport: Greenwood Press.

People for Open Space (1980) *Endangered Harvest: The Future of Bay Area Farmland*, San Francisco.

Perrin, N. (1978) *First Person Rural, Essays of a Sometime Farmer*, Boston: David R. Godine.

—— (1980) *Second Person Rural*, Boston: David R. Godine.

—— (1984) *Third Person Rural*, Boston: David R. Godine.

—— (1991) *Last Person Rural*, Boston: David R. Godine.

Perry, C., Gore, A. and Fleming, L. (1986) *Old English Villages*, London: Weidenfeld and Nicholson.

Pettifer, J. and Turner, N. (1984) *Automania*, Boston: Little, Brown and Co.

Pollard, S. (1959) *A History of Labour in Sheffield*, Liverpool: Liverpool University Press.

Pollard, S. and Crossley, D. (1968) *The Wealth of Britain, 1085–1966*, London: Batsford.

Priddle G. and Kreutwizer, R. (1977) 'Evaluating Cottage Environments', in J. T. Coppock (ed.) *Second Homes: Curse or Blessing?*, Oxford: Pergamon.

Prince, H. (1967) *Parks in England*, Isle of Wight: Pinhorns.

—— (1986) 'Parks and Open Spaces', in H. Clout and P. Wood (eds) *London: Problems of Change*, London: Longman.

Punter, J. (1974) 'Urbanites in the Countryside', Toronto: University of Toronto, unpublished Ph.D. thesis.

Ragatz, R. (1977) 'Vacation Homes in Rural Areas: Towards a Model for Predicting Their Distribution and Occupancy Patterns', in J. T. Coppock (ed.) *Second Homes: Curse or Blessing?*, Oxford: Pergamon.

Relph, E. (1976) *Place and Placelessness*, London: Pion.

—— (1981) *Rational Landscapes and Humanistic Geography*, London: Croom Helm.

—— (1988) *The Modern Urban Landscape*, London: Croom Helm.

Richards, J. M. (1973) *Castles on the Ground: The Anatomy of Suburbia*, London: John Murray.

Robinson, G. M. (1990) *Conflict and Change in the Countryside*, London: Belhaven.

Rohrer, W. and Douglas. L. (1969) *The Agrarian Transition in America: Dualism and Change*, New York: Bobbs-Merrill.

Rubinstein, D. and Speakman, C. (1969) *Leisure, Transport and the Countryside*, London: Fabin Research Series.

Runte, A. (1979) *National Parks: The American Experience*, Lincoln: University of Nebraska Press.

Ruskin, J. (1866) *The Crown of Wild Olive*, Philadelphia: H. Altemus, 1899 edition.

Saville, J. (1957) *Rural Depopulation in England and Wales, 1851–1951*, London: Dartington Hall Studies in Rural Sociology.

Schaffer, D. (1982) *Garden Cities for America: The Radburn Experience*, Philadelphia: Temple University Press.

Schmitt, P. (1969) *Back to Nature: The Arcadian Myth in Urban America*, New York: Oxford University Press.

Schuyler, D. (1986) *The New Urban Landscape*, Baltimore: Johns Hopkins University Press.

Sheail, J. (1976) *Nature in Trust: The History of Nature Conservation in Britain*, London: Blackie.

—— (1981) *Rural Conservation in Inter-War Britain*, London: Oxford University Press.

Shi, D. (1985) *The Simple Life: Plain Living and High Thinking in American Culture*, New York: Oxford University Press.

Shoard, M. (1985) *The Theft of the Countryside*, London: Temple Smith.

Short, J. R. (1991) *Imagined Country*, London: Routledge.

Simmons, I. G. (1975) *Rural Recreation in the Industrial World*, London: Edward Arnold.

Sorokin, P., Zimmerman, C. and Gilpin, C. (eds) (1965) *A Systematic Source Book in Rural Sociology*, New York: Russell and Russell.

Spectorsky, A. (1955) *The Exurbanites*, Philadelphia: Lippincott.

Stauffer, D. B. (1974) *A Short History of American Poetry*, New York: Dutton.

223

Stein, C. (1950) *Towards New Towns for America*, Cambridge, Mass.: MIT Press.
Steiner, G. (1971) *In Bluebeard's Castle*, London: Faber and Faber.
Stephenson, T. (1989) *Forbidden Land: The Struggle for Access to Mountain and Moorland*, Manchester: Manchester University Press.
Stern, A. (ed.) (1981) *The Anglo-American Suburb*, New York: St. Martin's Press.
Stilgoe, J. (1988) *Borderland: Origins of the American Suburb, 1820–1939*, New Haven: Yale University Press.
Stokes, S. (1989) *Saving America's Countryside: A Guide to Rural Conservation*, Baltimore: Johns Hopkins University Press/National Trust for Historic Preservation.
Stone, L. (1967) *Urban Development in Canada*, Ottawa: Dominion Bureau of Statistics.
Strong, D. (1988) *Dreamers and Defenders: American Conservationists*, Lincoln: University of Nebraska Press.
Taylor, N. (1973) *The Village in the City*, London: Temple Smith.
Thacker, C. (1983) *The Wildness Pleases: The Origins of Romanticism*, London: Croom Helm.
Thernstrom, S. (1973) *The Other Bostonians: Poverty and Progress in the American Metropolis, 1880–1970*, Cambridge: Harvard University Press.
Thomas, K. (1983) *Man and the Natural World: Changing Attitudes in England 1500–1800*, London: Allen Lane.
Thompson, E. (1968) *The Making of the English Working Class*, London: Penguin.
Thoreau, H. (1854) *Walden, or Life in the Woods*, New York: New American Library, 1960 edition.
Treble, R. (1989) 'The Victorian Picture of the Country', in G. E. Mingay (ed.) *The Rural Idyll*, London: Routledge.
Tuan, Y-F. (1974) *Topophilia*, New York: Prentice-Hall.
Tunbridge, J. E. (1981) 'Conservation Trusts as Geographic Agents: Their Impacts Upon Landscape, Townscape and Land Use', *Transactions, Institute of British Geographers*, New Series 6: 104–25.
Turner, F. J. (1920) *The Frontier in American History*, New York: Holt.
United States Dept of Agriculture (1962) *Report of the Outdoor Recreation Resources Review Commission*, Washington, DC.
Unwin, R. (1981) *Nothing Gained By Overcrowding*, London: Garden Cities and Town Planning Association.
Walvin, J. (1978) *Leisure and Society, 1830–1950*, London: Longman.
Ward, D. (1971) *Cities and Immigrants: A Geography of Change in Nineteenth Century America*, New York: Oxford University Press.
Warkentin, J. (1966) 'Southern Ontario: A View from the West', *The Canadian Geographer*, 10: 157–71.
Webb, B. and Webb, S. (1923) *The Decay of Capitalist Civilization*, New York: Harcourt, Brace.
Weber, M. (1899) *The Growth of Cities in the Nineteenth Century: A Study in Statistics*, Ithaca: Cornell University Press.
Wheeler, W. (1976) 'Jeffersonian Thought in Urban Society', in J. Marth (ed.) *The Agrarian Tradition*, Knoxville: University of Tennessee.
Whitaker, B and Browne, K. (1971) *Parks for People*, New York: Winchester Press.
White, K. (1977) *Country Life in Classical Times*, London: Paul Elek.
White, M. and White, L. (1962) *The Intellectual Versus the City*, Boston: Harvard University Press.
Whyte, W. (1956) *The Organization Man*, New York: Doubleday.

Wiener, M. (1981) *English Culture and the Decline of the Industrial Spirit*, New York: Cambridge University Press.

Wildlife Link (1990) *Annual Report, 1989*, London.

Williams, G. (1978) *The Royal Parks of London*, London: Constable.

Williams, M. (1972) *Thomas Hardy and Rural England*, London: Macmillan.

Williams, R. (1960) *Culture and Society*, London: Chatto and Windus.

—— (1966) 'Introduction' in R. Jefferies, *Hodge and His Masters*, London: MacGibbon and Kee.

—— (1973) *The Country and the City*, London: Chatto and Windus.

Williams-Ellis, C. (1928) *England and the Octopus*, London: G. Bles.

Wilson, A. (1990) *The Culture of Nature*, Cambridge, Mass.: Blackwell.

Wolfe, R. I. (1965) 'About Cottages and Cottagers', *Landscape*, 15: 6–8.

Wood, J. (1982) 'Village and Community in Early Colonial New England', *Journal of Historical Geography* 8: 333–46.

Woodruffe, B. (1982) *Wiltshire Villages*, London: Robert Hale.

Woodward, G. (1865) *Woodward's Country Homes*, New York: Woodward.

Wright, F. Lloyd (1958) *The Living City*, New York: Horizon Press.

225

INDEX